Toughening of Plastics

ACS SYMPOSIUM SERIES **759**

Toughening of Plastics
Advances in Modeling and Experiments

Raymond A. Pearson, EDITOR
Lehigh University

H.-J. Sue, EDITOR
Texas A&M University

A. F. Yee, EDITOR
The University of Michigan

American Chemical Society, Washington DC

Library of Congress Cataloging-in-Publication Data

Toughening of Plastics : testing methods and advances in modeling and experiments / Raymond A. Pearson, editor, H.-J. Sue, Editor, A.F. Yee, editor.

    p.   cm.—(ACS symposium series , ISSN 0097–6156 ; 759)

    Includes bibliographical references and index.

    ISBN 0–8412–3657–7

    1. Polymers—Additives—Congresses. 2. Polymers—Mechanical properties—Congresses.

    I. Pearson, Raymond A., 1958– . II. Sue, H.-J. (Hung-Jue, 1958– . III. Yee, A. F., 1945– . IV. Series.

TP1142 .T68   2000
668.4´11—dc21   00–26260

The paper used in this publication meets the minimum requirements of American National Standard for Information Sciences—Permanence of Paper for Printed Library Materials, ANSI Z39.48–1984.

Foreword

THE ACS SYMPOSIUM SERIES was first published in 1974 to provide a mechanism for publishing symposia quickly in book form. The purpose of the series is to publish timely, comprehensive books developed from ACS sponsored symposia based on current scientific research. Occasionally, books are developed from symposia sponsored by other organizations when the topic is of keen interest to the chemistry audience.

Before agreeing to publish a book, the proposed table of contents is reviewed for appropriate and comprehensive coverage and for interest to the audience. Some papers may be excluded in order to better focus the book; others may be added to provide comprehensiveness. When appropriate, overview or introductory chapters are added. Drafts of chapters are peer-reviewed prior to final acceptance or rejection, and manuscripts are prepared in camera-ready format.

As a rule, only original research papers and original review papers are included in the volumes. Verbatim reproductions of previously published papers are not accepted.

ACS BOOKS DEPARTMENT

Contents

INDEXES

Preface

It is generally accepted that polymers without additives would be commercial failures. The toughening of polymers involves the use of additives called toughening agents. The purpose of toughening agents is to impart impact resistance and to increase damage tolerance. Typically, elastomeric polymers are used to toughen engineering plastics although rigid polymers and inorganic fillers can also be used as toughening agents.

The applications of toughened polymers are commonplace today. Toughened plastics can be found in the cars we drive, in the appliances in our home, in the electronic devices in our offices, and even in some of the sports equipment that we use on weekends. The gains made in understanding the fundamentals of toughening have resulted in new products and new applications. In summary, the technology of toughening plastics has had a very positive effect on society.

All of the chapters in this book are based on the papers presented at the Fall 1998 American Chemical Society (ACS) Meeting in Boston, Massachusetts and their abstracts can be found in Volume 79 of the Proceedings of the American Chemical Society: Division of Polymeric Materials: Science and Engineering, Inc. The abstracts for the Toughening of Plastics Symposium can be found on pages 140 to 217, which contain over 30 contributions. Unfortunately, these papers are rather brief, hence our motivation for publishing a more comprehensive compilation of papers.

The papers in this book were chosen because of their significance to the technology of toughened polymers. Many of the papers are from leading worldwide experts. We have carefully selected 12 papers that seem to capture the excitement of the symposium. These papers represent the most recent advances in synthesis, processing, characterizing, and modeling the toughening of plastics. It is our belief that the compilation of the modeling efforts has more depth than any publication to date.

Acknowledgments

The ACS has tracked this important technology for many years and has organized symposia on "Rubber-Modified Thermoset Resins," "Rubber-Toughened Plastics," "Toughened Plastics I," and "Toughened Plastics II".

Therefore, we are indebted to all contributors for their time and patience. We also thank the ACS Division of Polymeric Materials: Science and Engineering, Inc. for sponsoring the workshop, the symposium, and this book.

RAYMOND A. PEARSON
Department of Materials Science and Engineering
Lehigh University
Bethlehem, PA 18015–3195

H.-J. SUE
Department of Mechanical Engineering
Texas A&M University
College Station, TX 77843–3123

A. F. YEE
Department of Materials Science and Engineering
University of Michigan
Ann Arbor, MI 48109

Chapter 1

Introduction to the Toughening of Polymers

Raymond A. Pearson

Department of Materials Science and Engineering,
Lehigh University, Bethlehem, PA 18015–3195

The purpose of this chapter is to provide the background needed for subsequent chapters on toughened plastics. This chapter introduces the topic of fracture mechanics, describes the methods used to quantify toughness, discusses the origin of toughness in terms of toughening mechanisms, briefly reviews the factors influencing toughness, and comments on the latest advances in the science of the toughening of polymers. The main focus is on polymers blends toughened by a soft rubber phase although several alternative approaches are briefly mentioned. The purpose of this book is to present the latest advances in the field of toughened polymers so that the materials specialist can use these latest concepts to create new materials with improved toughness.

Introduction

The purpose of this chapter is to provide an introductory review of toughened polymers. In this section, a brief review of the development of polymer blends is attempted. It will be revealed that the first toughened polymer blends were binary mixtures of rubber and plastic, the so called rubber-toughened plastics. Ternary blends were developed soon after the simple binary mixtures and also contain a rubbery phase for toughness. Most recently, tough rigid-rigid polymer alloys have been developed in an attempt to improve toughness without sacrificing strength.

The history of polymer blends and alloys is described in detail in a book by Utracki [1]. For introductory purposes a brief description of polymer blends is contained below.

The first modern thermoplastic blend is often identified as a PVC/NBR blend which, in early 1942, NBR was discovered to permanently plasticize PVC. A co-polymerized polystyrene-polybutadiene blend was introduced by Dow later that same year. Soon after the introduction of PS-PB blends, mechanical mixtures of NBR and SAN were developed thus ABS blends were born. ABS-type blends dominate the blend market and, on average, 80 new blends of ABS are introduced to the US market place each year [1].

Toughened engineering polymer blends were developed as early as 1960 when it was discovered that polystyrene (PS) was miscible with poly-2,6-dimethyl-1,4-polyphenyleneether (PPE). Polystyrene-butadiene copolymers were added to improve impact resistance. These ternary blends, known in the industry as the Noryl™ blends, were commercialized by the General Electric company. A number of other toughened engineering polymer blends soon followed including super tough nylon, rubber-toughened PBT/PET, and rubber-toughened polycarbonate (PC). Ternary blends of PC/PBT/ rubber and PPE/ PA/ rubber were also commercialized. The most frequently claimed property contained in patents during this period was high impact strength.

In the early 1980s it was discovered that epoxy resins could be toughened by the addition of a rigid thermoplastic phase, thus the rigid-rigid polymer alloy concept was born [2]. The advantage of using a rigid thermoplastic phase over a soft rubbery phase is that there is not a drop in modulus or strength. Such behavior is critical for matrices used in advanced composites. Polyether sulfone, poly sulfone, polyetherimide, polyphenylene ether, and polybutylene terephthalate have been evaluated as toughening agents for a wide variety of thermosetting plastics [3] and in some cases, significant increases in toughness has been achieved.

Of course, the use of inorganic fillers to modify the properties of polymers has been around for many years. However, much of the focus on filled-polymers has been on the reinforcing effect, yet it has long been recognized that fracture toughness is also improved [4,5]. Perhaps the lack of attention given to inorganic fillers as toughening agents is due to their poor response to impact situations and the sharp surfaces generated by an impact event. However, the fact remains that the addition of inorganic spheres, platelets or short fibers can improve the static fracture toughness of many polymers.

The main reasons for blending, compounding and reinforcing plastics are economy and performance. If a material can be generated that will lower the overall cost while maintaining or improving performance of a particular product then the manufacturer must use it to remain competitive. The purpose of this book is to present the latest advances in the field of toughened polymers so that the materials specialist can use these latest concepts to create new materials with improved toughness.

Measurements of Toughness

The toughness of a material can be measured using a variety of techniques. Tensile testing (ASTM D638) is perhaps the simplest technique. The area under the stress strain curve is often used to quantify toughness. However, the mechanical behavior of polymers is extremely rate and stress-state dependent [6]. Therefore, more complicated tests have been derived to predict the performance of plastic products. For example, the ASTM D 3763 puncture test and the ASTM D256 Izod test utilize biaxial and triaxial stress states at impact velocities. Unfortunately, the values provided by these impact tests cannot be used in directly in design. Preliminary screening of materials can be accomplished by looking at ductile-brittle transition temperatures of the appropriate impact test. The qualitative nature of the impact tests make them better-suited for quality control purposes than for design For quantitative measures of toughness, most researchers employ the use of fracture mechanics.

Linear Elastic Fracture Mechanics (LEFM) is now widely used to quantify the toughness of polymers (ASTM D5045). Two test geometries are covered by the ASTM test method for plane-strain fracture toughness and strain energy release rate of plastic materials. See Figure 1. Fracture toughness is often quantified using a stress intensity approach but strain energy release rates can also be used. There are numerous textbooks on fracture toughness testing and it is recommended to review the books written by Broek [7], Hertzberg [8], and Kinloch and Young [9].

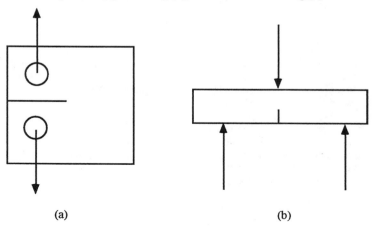

(a)   (b)

Figure 1: *Schematic diagrams of the a) compact tension (CT) and b) Single-Edge-Notched, 3-Point Bend (SEN-3PB) specimens, which are often used to determine the fracture toughness of polymers.*

The stress intensity factor, K, is a parameter used to relate the applied stress, σ, and the flaw size, a:

$$K = \sigma Y \sqrt{a} \qquad (1)$$

Where Y is a shape factor. Note that the derivation of K assumes an infinitely cracked plate and the shape factor takes into account edge effects in specimens of finite size. The subscripts I, II, and III are added to K to denote crack opening, crack shearing or crack twisting modes, respectively. Mode I or the crack opening mode represents a lower limit and is useful in engineering design and material selection. Note that K is a parameter to quantify the stress field around the crack tip and K_C is the critical stress intensity factor, which is a material parameter used to predict crack propagation. For mode I loading situations, the crack will propagate when $K_I > K_{IC}$.

G is a measure of the amount of energy available to extend the crack (per unit area), i.e. $G = dU/da$. Where U is strain energy available and a is the crack length. Like the stress intensity concept, G_I is a parameter that quantifies the amount of energy available in a mode I situation and G_{IC} is the critical strain energy release rate for crack propagation, which is a material property. The same specimens used to determine K_{IC} can be used for G_{IC}. The following equation is used:

$$G_{IC} = \frac{U}{BW\phi} \tag{2}$$

Where B is the specimen thickness, W is the specimen width, and ϕ is the energy calibration factor for the type of specimen used. G can also be determine from K by

$$G_{IC} = \frac{(1 - v^2)K_{IC}^2}{E} \tag{3}$$

Where v is Poisson's ratio and E is the Young's modulus. Since the mechanical properties of polymers are time and temperature dependent, it is preferred to determine G directly. For a more detailed description of the fracture test, the careful reader is referred to ASTM D 5045 test method.

For ductile materials, the measured strain energy release rate is related to the elastic energy needed to separate the two surfaces plus the plastic energy consumed during separation. The plastic energy clearly dominates, thus the magnitude of G_{IC} is related to the amount of plastic deformation present at the crack tip. An estimation of the plastic zone size can be made using the equation proposed by Irwin:

$$r_p \approx \frac{1}{6\pi} \frac{K_{IC}^2}{\sigma_{ys}^2} \tag{4}$$

Where r_p is the plastic zone radius, and σ_{ys} is the yield strength of the material. Note that materials are often toughened by the use of additives that increase the size of the plastic zone and, hence, increase the amount of the plastic energy consumed in propagating the crack.

Note that LEFM is only valid when

$$B \text{ and } a \ge 2.5\left(\frac{K_{IC}}{\sigma_{ys}}\right)^2 \tag{5}$$

Where B and a are specimen thickness and crack length, respectively.

For polymers possessing high toughness and low yield strength, it is impractical, if not impossible, to make specimens that properly constrain the crack. Therefore, LEFM characterizations cannot be used. Fortunately, there are several Elastic Plastic Fracture Mechanics (EPFM) approaches that can be used to determine the fracture toughness of these fracture resistant polymers. We will focus our attention on the J-integral approach but the careful reader should be aware that concepts such as "essential work of fracture" are also under development [7,8].

The J-integral concept defines J as a path-independent integral that describes the stresses and strains around the crack tip for linear or nonlinear elastic deformations. For the 2D case, J integral is defined by

$$J = \int_C \left(Wdy - T \cdot \frac{\partial u}{\partial x} ds \right) \tag{6}$$

where x, y are the rectangular coordinates perpendicular to the crack front, ds is the increment along contour C, T is the stress vector acting on the contour, u is the displacement vector, and W is the strain energy density (= $\int \sigma d\varepsilon$). It is necessary to assume that the crack tip plastic deformation that occurs under monotonic loading (without unloading) is identical to that of non-linear elastic deformation, this has been shown to be true for linear elastic materials:

$$J_{IC} \text{ (plastic test)} = G_{IC} \text{ (elastic test)} \tag{7}$$

Landes and Begely [10] developed a simple method for the determination of the plastic component of J_{IC} that involves a plate containing a deep notch subjected to pure bending. It serves as the basis for the ASTM standard that uses three point bend specimens with a span to width ratio of four. J is given as

$$J_I = \frac{K_I^2}{E}\left(1 - v^2\right) + \frac{2A}{Bb} \tag{8}$$

Where A is the area under the stress-strain curve and b is the length of the unbroken ligament.

In summary, three point bending and compact tension can be used to determine the fracture toughness of polymers. LEFM type of characterizations can be used when the plastic deformation at the crack tip can be properly constrained. Very tough polymers must be characterized using EPFM in order to obtain a parameter that is geometry independent and useful for design. In the next section, we will explore the techniques used to study the plastic deformation mechanisms responsible for the high toughness values.

Fractography and Fracture Mechanisms

The study of plastic deformation mechanisms occurring at crack tip often involves the use of both electron and optical microscopes. Scanning electron microscopes are used to inspect fracture surfaces, the regions of highest stress and deformation. Transmission electron and optical microscopy are used to search for damage that occurs below the surface. Since transmission microscopy investigations involve extensive specimen preparation, such techniques are not as popular as SEM.

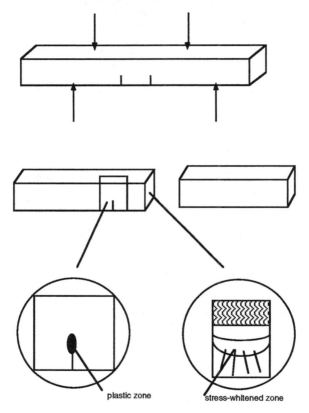

plastic zone   stress-whitened zone

Figure 2: A schematic diagram of a four-point bend double crack specimen that can be used to examine both surface and subsurface damage. This technique facilitates the loading of a crack to its maximum K without unstable crack growth.

Figure 2 contains a schematic diagram of a four-point bend double crack specimen that can be used to examine both surface and subsurface damage in toughened polymers [11]. The advantage of the 4PB-DN specimen is that one crack will reach its critical K and propagate thereby unloading the second that has been loaded near its maximum K. Scanning electron microscopy can be used to inspect deformation mechanisms on fractured surfaces [12]. Usually, the surface is coated with gold to reduce charge build-up that often occurs when imaging from secondary electrons. Occasionally, surfaces are stained with osmium tetroxide to facilitate the contrast of phases in polymer blends. Note that backscattered electrons are used to form these images and the stained regions appear bright. Transmission miscroscopy can be used to look at subsurface damage [13]. In the case of TEM, thin sections near the crack tip are taken using cryogenic microtoming techniques. The rubber phase is stained with osmium tetroxide to provide phase contrast and appear dark. TEM is useful for studying crazing but shear banding cannot be detected directly in a TEM. Therefore TOM, which uses a light microscope, is often used with crossed polars to detect regions of permanent orientation, i.e. plasticity [14].

Examples of a micrograph taken from fracture surface of a rubber toughened epoxy is shown in Figure 3a. Note the appearance of "holes" on the fracture surface. These "holes are actually lined with rubber that has cavitated under triaxial stress. Also, note the matrix dilation around the cavitated rubber particles. Figure 3b contains a transmission optical micrographs of a petrographically polished thin section that reveals the subsurface damage. Note the formation of a large damage zone at the crack tip is expected for a toughened polymer. Not shown is the formation of shear banding in the matrix. Rubber particle cavitation and concomitant matrix shear banding is a common toughening mechanism in engineering polymers toughened with rubber particles.

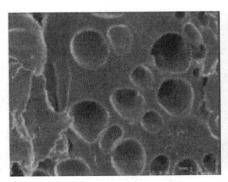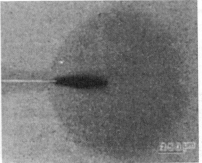

Figure 3: a) Fracture surface of rubber-toughened epoxy and b) subsurface damage in a rubber-toughened epoxy. Reproduced with permission from reference 24.

Figure 4 contains schematic diagrams of the two types of toughening mechanisms found in rubber toughened polymers: massive crazing and massive shear banding. Massive crazing is seen in blends where the matrix prefers to craze such as high impact polystyrene (HIPS) [15-17]. Massive shear banding has been observed in rubber-toughened polycarbonate]18-20] and rubber-toughened epoxies [21-26]. There are also a number of polymers such as ABS [27] and rubber-toughened PMMA [28-30] that exhibit both shear banding and crazing at the crack tip.

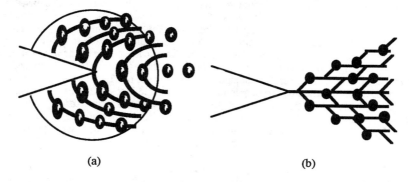

(a)   (b)

Figure 4: Schematic diagrams of idealized crack-tip plastic zones: a) massive shear banding and b) massive crazing.

Factors Influencing Toughness

There are a number of factors that influence the amount of toughening obtained by the use of additives and fillers. Most of these factors have been experimentally determined, however, recent mechanical modeling has improved our understanding of these important toughening mechanisms.

In rubber-toughened plastics, the matrix plays an important role in determining the overall toughness. Some matrices tend to craze because of low entanglement density[31]. Massive crazing induced by rubber particles is clearly observed in high-impact polystyrene. In crazing polymers, high molecular weight is needed to stabilize crazes. Highly entangled polymers tend to deform via shear banding. High molecular weight matrices are, in general, tougher than their low-molecular counterparts. Shear banding is clearly observed in lightly crosslinked epoxies. In the case of epoxies, lower crosslink density produces a more "toughenable" polymer upon rubber addition.

Rubber particle size has been studies by a number of investigators [32-34]. Rubber particles over 5 microns in diameter are often too large to interact with the stress field at the crack tip. Rubber particles under 100 nm in diameter appear to be too small to cavitate effectively. Without the cavitation of the rubber particles.

subsequent matrix shear banding in the presence of a triaxial stress field at the crack tip is very unlikely.

The effect of rubber concentration on toughness has been studied by a large number of researchers [35-40]. Some have attributed a minimum amount of rubber needed to an interparticle distance effect (mostly semicrystalline plastics while others have shown a linear increase in toughness with rubber content. At high rubber content the toughness decreases presumably due to the fact that there is less matrix available for massive shear banding or crazing.

The type of rubber has been shown to be important by a number of researchers [41-43]. For example, in nylon the type of rubber has been shown to be more important than interparticle distance. The effect of the type of rubber has been often associated with the caviational strength of the rubber particle. However, it is important to note that the blend morphology and matrix-particle interphase regions may also change.

Rubber particle strength has also been examined [44-46]. A number of investigators have shown that epoxies containing microvoids are as tough their rubber-modified counterparts. However, these micro-voided materials are much more difficult to process than conventional rubber-toughened polymers and the microvoid modification has only been applied to a few number of matrices. Therefore, the utility of the microvoided polymer concept is limited however, the study of these systems do challenge our understanding of the role of rubber-particle cavitation resistance.

The role of the rubber particle–matrix interface has been studied by a few groups with mixed results [47-49]. Some researchers have found that strengthening the interface improves toughness, others have shown that a diffuse interphase region improves toughness, while others have shown that the interphase region can control the amount of matrix dilation around the rubber particles.

The role of the particle morphology has been studied by a number of researchers [50-51] Bucknall et al. [50] have studied the role of a salami type microstructure in rubber-toughened polystyrene. Lovell et al. [51] have studied the effect of multilayered core-shell latex particles on toughening PMMA. There appears to be advantages over the simple on phase particles.

The role of blend morphology in rubber-toughened plastics has also been studied [52-53]. Clearly, the rubber phase has to be uniformly distributed for improved toughness. However, there may be an advantage of having segregation on a microscopic scale as shown by Bagheri et al. [52] and by Qian et al. [53].

In summary, there are a number of variables to consider when developing toughened plastics and their interrelationships remain an area of active study.

Key Issues and Questions

Experimental studies have provided a wealth of information but sometimes the data contradicts itself or too many parameters are changed at the same time. Therefore, much emphasis has been placed on our ability to model rubber toughened systems. Proper mechanical modeling should help us interpret the experimental studies on toughened plastics. A number of modeling studies have been performed and are briefly summarized below.

Broughtman and Panzzia [54] where perhaps the first researchers to use Finite Element Methods to facilitate our understanding of rubber-toughened plastics. Their model emphasized the overlap of stress fields associated with the rubber particles and eluded to the interaction of the particles. Wu took the particle interaction concept one step further and proposed a critical interparticle distance concept as a matrix dependent parameter that controls 'toughenability'. Unfortunately, the critical interparticle distance model has been scrutinized by many researchers and its utility appears doubtful. However, the discussion on the validity of the interparticle distance model has focused attention to the cavitational strength of the rubber particles. Indeed much progress has been made in understanding the role of rubber particle cavitation in toughening mechanisms.

A number of FEM models have been developed since the pioneering work of Broughtman and Panizzia. For example, several researchers [55-57] have applied the FEM appraoch to examine the role of rubber particle cavitation in shear band formation at crack tips. Sue et al. [58] have used FE to show that rigid organic particles could also intiate shear banding when under an uniaxial tension stress field. The use of rigid organic particles to promote massive shear banding at crack tips appears unlikely due to the lack of particle cavitation. A significant outcome in the past few years is the recognition that particle cavitation is a critical stage in the development of shear bands at crack tips.

In spite of the enormous progress made in the last 20 years on toughened plastics there remains a number of key questions:

1) What parameters control the size and shape of the plastic zone?
2) What is happening at the molecular level? Chain slippage versus chain scission.
3) What is the optimal particle size?
4) What is the optimal blend morphology? Uniforn particle distribution versus microclustered paticulate morphologies.
5) Can we design toughened polymers with multiple toughening mechanism that interact in a synergistic fashion?
6) Can rigid organic particles ever replace rubber-toughening?
7) Are there new chemical routes for producing toughened plastics?

Some of these questions may never be answered but the chapers in this book represent the latest advances in the field of toughened plastics.

References

1. Utracki, L.A., Polymer Alloys and Bends: Thermodynamics and Rheology, Hanser Publishers: New York, 1989.
2. Bucknell, C. B.; Partridge, I. K. *Polymer* **1983**, 24, 639.
3. Pearson, R.A., in Toughened Plastics I: Science and Engineering,edited by C. K. Riew and A. J. KinlochAmerican Chemical Society: Washington, D. C. 1993, pp. 405-425.
4. Spanoudakis, L.; Young, R. J. *J. Mater. Sci.* **1984**, 19, 473.
5. Maloney, A. C.; Kausch, H. H.; Kiaser, T.; Beer, H. R. *J. Mater. Sci.* **1987**, 22, 381.
6. Ward, I. M., Mechanical Properties of Solid Polymers 2nd Ed., John Wiley and Sons: New York 1985, pp. 329-398.
7. Broek, D., Elementary Engineering Fracture Mechanics, Kluwer: Dordrecht 1991.
8. Hertzberg, R. W. Deformation and Fracture Mechanics of Engineering Materials 4th Ed. John Wiley and Sons: New York 1996, pp 211-374.
9. Kinloch, A. J. ; Young, Fracture Behavior of Polymers, Applied Science Publishers: London 1983.
10. Landes, J. D.; Begley, J. A., *ASTM STP 632* **1977**, 57.
11. Sue, H. J.; Pearson, R. A.; Parker, D. S.; Huang, J.; Yee, A. F. *in ACS Div. Polym. Chem. Polym. Prep.* **1998**, 29, 147.
12. A. F. Yee and R. A. Pearson, Chapter 8 in Fractography: and Failure Mechanisms of Polymers and Composites, Editor: Anne C. Roulin-Maloney, Elsevier Science: New York, 1989, 291.
13. A.S. Holik; Kambour, R. P.; Hobbs, S. Y.; Fink, D. G., *Microstruct. Sci.* **1979**, 7, 357.
14. Sue, H. J.; Garcia-Meitin; Pickelman, D. M.; Yang, P. C. in Toughened Plastics I: Science and Engineering,edited by C. K. Riew and A. J. Kinloch, American Chemical Society: Washington, D. C. 1993, pp. 259-291.
15. Bucknall, C. B.; Smith, R. R. *Polymer* 1965, 6, 471.
16. Hobbs, S. Y. *Polym. Eng. Sci.* **1986**, 26, 74.
17. Keskkula, H. in ACS: *Adv. In Chem. Ser.* **1989**, 222, 289-299.
18. Parker, D. S.; Sue, H.-J.; Huang, J; Yee A. F. *Polymer* **1990**, 31, 2267.
19. Cheng, T. W.; Keskkula; Paul, D. R. J. *Appl. Polym. Sci.* **1992**, 45, 1245.
20. Lee, C.-B.; Chang, F.-C. *Polym. Eng. Sci.* **1992**, 32, 792.
21. Kinloch, A. J.,;Shaw, S. J.;Tod, D. A.; Hunston, D. L. *Polymer* **1983**, 24, 1341.
22. *Idem, Ibid,* 1355.
23. Pearson, R. A.; Yee, A. F. *J. Mater. Sci.* **1986**, 21, 2475.
24. Pearson, R. A.; Yee, A. F. *J. Mater. Sci.* **1989**, 24, 2571.
25. Garg, A. C.; Mai, Y. W. *Compos. Sci. Technol.* **1988**, 31, 179.
26. Kozii, V.V.; Rozenburg, B. A. *Polym. Sci.* **1992**, 34, 919.
27. Morbitzer, L.; Kranz, D.; Humme, G.; Ott, K. H. *J. Appl. Polym. Sci.* **1976**, 20, 2691.
28. Hooley, C. J.; Moore, D. R.; Whale, M.; Williams, M. J. *Plastics and Rubber Processing and Applications* **1981**, 1, 345.
29. C. Wrotecki,; Hiem, P.; Gaillard, P. *Polym. Eng. Sci.* **1991**, 31, 213.

30. Nelliappan, V., El-Aasser, M.S., Klein, A.; Daniels, E.S.; Roberts, J. E.: Pearson, R. A. *J. Appl. Polym. Sci.* **1997**, 65, 581.
31. Kramer, E. J. Chapter 1 in Advances in Polymer Science, Springer-Verlag: Berlin 1983, vol. 52/53, pp.1-56.
32. Pearson, R. A.; Yee, A. F. *J. Mater. Sci.* **1991**, 26, 3828.
33. Cigna, G.; Lomellini, P.; Merlotti, M. *J. Appl. Polym. Sci.* **1989**, 37, 1540.
34. Azimi, H. R.; Pearson, R. A.; Hertzberg, R. W. *J. Mater. Sci* 1996, 31, 3777.
35. Yee, A. F. ; Pearson, R. A. *J. Mater. Sci.* **1986**, 21, 2462.
36. Borggreve, R. J. M., Gaymans, R. J.; Schuijer; Ingen Houz, F. J. *Polymer* **1987**, 28, 1489.
37. Hobbs, S. Y.; Bopp, R. C.; Watkins, V. H. *Polym. Eng. Sci.* **1983**, 380.
38. Wu, S. *J. Appl. Polym. Sci.* **1988**, 35, 549.
39. Bucknall, C.B.; Cote, F.F.; Partridge, I. K. *J. Mater. Sci.* **1986**, 21, 301.
40. Muratoglu, O. K.; Argon, A.S.; Cohen, R. E.; Weinburg, M. *Polymer* **1995**, 36, 921.
41. Borggreve, R. J. M.; Gaymans, R. J. ; Schuijer, J. *Polymer* **1989**, 30, 71.
42. Dompas, D.; Groeninckx, G. *Polymer* **1994**, 35, 4743.
43. Laurienzo, P.; Malinconico, M; Martucelli, E.: Volpe, M. G. *Polymer* **1991**, 30, 835.
44. Borggreve, R. J. M.; Gaymans, R. J. ; Eichenwald, H. M. *Polymer* **1989**, 30, 78.
45. Bagheri, R.; Pearson, R. A. *Polymer* **1996**, 20, 4529.
46. Lazzeri, A.; Bucknall, C. B.; *Polymer* **1995**, 36, 2895.
47. Bagheri;, R.; Pearson, R. A. *J. Appl. Polym. Sci.* **1995**, 58, 427.

48. Huang, Y.; Kinloch, A. J.; Bertsch, R.; Sibert, A. R. Advances in Chemistry Series,Edited by C. KJ. Riew and A. J. Kinloch, ACS: Washington, D. C. 1993, 233, 189.
49. Chen, T. K.; Jan, Y. H. *Polym. Eng. Sci.* **1991**, 31, 577.
50. Lovell, P. A., McDonald, J.; Saunders, D. E. J.; Young, R. J. *Polymer* **1993**, 34, 61.
51. Bucknall, C. B.; Partridge, I. K. in <u>Toughening of Plastics II</u>, PRI:London 1991, 28/1.
52. Bagheri, R.; Pearson, R. A. *J. Mater. Sci.* **1996**, 31, 3945.
53. Qian, J. Y.; Pearson, R. A.; Dimonie, V. L.; Shaffer, O. L.; El-Aasser, M. S. *Polymer* **1997**, 38, 21.
54. Broutman, L. J.; Panizza, G. *Int. J. Polym. Mater.* 1971, 1, 95.
55. Huang, Y.; Kinloch, A. J. *Polymer* **1992**, 33, 1330.
56. Kinloch, A. J.; Guild F. J. Toughened Plastics II: Novel Approaches in Science and Engineering, edited by C. K. Riew and A. J. Kinloch, ACS: Washington D. C. 1996, p. 1.
57. Fukui, T.; Kituchi, Y; Inoue, T. *Polymer* **1991**, 32, 2367.
58. Sue, H. J.; Pearson, R. A.; Yee, A. F. *Polym. Eng. Sci.* **1991**, 31 793.

MODELING STUDIES

Chapter 2

Recent Developments in the Modeling of Dilatational Yielding in Toughened Plastics

Andrea Lazzeri[1,3] and Clive B. Bucknall[2]

[1]University of Pisa, 56126 Pisa, Italy
[2]SIMS B61 Cranfield University, Cranfield,
Bedford MK43 OAL, United Kingdom

A quantitative model for cavitation and consequent dilatational yielding in multiphase plastics (*1*, *2*) is reviewed and new developments are reported and compared with experimental results. According to the model, cavitation can occur by debonding at phase boundaries or by nucleation of voids within a soft polymeric phase when the stored volumetric strain energy density within the rubber phase exceeds a critical value. The model relates the critical volume strain required for cavitation to the properties of the particle: its size, shear modulus, surface energy and failure strain in biaxial extension. Subsequent to cavitation of the rubber particles, the yield behavior of the polymer is significantly altered, especially at high triaxiality, and can be modeled by the modified Gurson equation proposed by Lazzeri and Bucknall (*1*, *2*). Particle cavitation also increases the rate of yielding.

On increasing the strain level in the material, the deformation will tend to assume an inhomogeneous character leading to the formation of dilatational bands. The particles located in proximity of a dilatational band will cavitate preferentially due to the hydrostatic stress concentration near its ends, leading to a propagation of the band. At the same time, the number of dilatational bands will increase and their growth is associated with significant levels of energy absorption. In some semi-crystalline polymers this can lead to the formation of microfibrils in the ligaments between neighboring particles.

Many commercial plastics are not simple single-component materials, but blends containing two or more plastics, or plastics with a rubbery phase. These "multiphase plastics" offer advantages in properties, performance and economics over comparable

---

[*] Current address: Massachusetts Institute of Technology, 77 Massachusetts Avenue, Cambridge (MA), 02139 USA.

single-component plastics and represent a substantial fraction of the world's total plastic consumption.

The fracture behavior of multiphase plastics is complex. Some general principles have been established, but until quite recently it has not been possible to develop quantitative relationships between structure and properties. Both composition and morphology are known to have a strong influence upon toughness, but experimental studies have revealed some puzzling trends and unexplained effects. Consequently, it has been difficult to design new multiphase plastics using established scientific principles.

This picture has been changing over the past few years, following developments in understanding the mechanisms of energy absorption in two-phase polymers, and in particular the introduction of a quantitative model for rubber particle cavitation and consequent dilatational yielding in multiphase plastics (1, 2). Cavitation can occur by debonding at phase boundaries or by nucleation of voids within a soft polymeric phase when the stored volumetric strain energy within the rubber phase exceeds a critical value. The proposed model relates the critical volume strain required for cavitation to the properties of the particle: its size, shear modulus, surface energy and failure strain in biaxial extension. Subsequent to cavitation of the rubber particles, the model predicts the formation of dilatational bands. This term describes planar yield zones that combine in-plane shear with dilatational extension in the direction normal to the plane. Several studies confirmed the formation of dilatational bands by means of small angle X-ray scattering (SAXS) and by light scattering. The band orientation with respect to the major tensile axis was also found to be in good agreement with model predictions.

New developments in the modeling of dilatational yielding in rubber toughened polymers will be presented. In particular it will be shown that cavitation of the rubber particles causes acceleration in the rate of shear yielding of the material.

Cavitation Model

In this section, the model for cavitation of rubber particles proposed by Lazzeri and Bucknall (1-3) will be briefly reviewed and discussed in relation to some experimental data.

It is assumed that rubber particles cavitate when, under positive hydrostatic stress, the stored volumetric strain energy density W_o:

$$W_o = \frac{K_r \Delta_V^2}{2} \tag{1}$$

within the rubber phase exceeds a critical value. In eq 1 K_r is the bulk modulus of the rubber and Δ_V is the volume strain of the particle. The energy density stored in the particle after the formation of a void of diameter d is then given by:

$$W(d) = \frac{K_r}{2}\left(\Delta_V - \frac{d^3}{D^3}\right)^2 + \frac{6\Gamma}{D}\frac{d^2}{D^2} + \frac{3G_r F(\lambda_f)}{2}\frac{d^3}{D^3} \tag{2}$$

where Γ is the surface energy of the rubber, D the particle diameter, G_r the rubber shear modulus and $F(\lambda_f)$ is a function of the elongation at break of the rubber in a biaxial state of stress. A necessary condition for cavitation is that the energy $W(d)$ has to be smaller than the initial energy before cavitation, W_o. At low volume strains, the energy $W(d)$ increases monotonically with d so that $W(d)$ is always greater than W_o. At high volume strains, however, $W(d)$ is no longer monotonic. The energy density first increases up to maximum (the "energy barrier" discussed below) and then decreases towards a minimum at which $W(d) < W_o$., i.e. the volumetric strain energy density of a particle with a hole of radius d is smaller than for an full rubber particle.

By imposing the conditions that both the first and the second derivative of W, with respect to d, are equal to zero, the critical volume deformation Δ_{VC} above which a cavitated particle is more stable than an intact particle, can be determined as:

$$\Delta_{VC} = 4\left(\frac{4\Gamma}{3K_r D}\right)^{3/4} + \frac{3G_r F(\lambda_f)}{2K_r} \tag{3}$$

As shown in Figure 1, cavitation occurs at a critical volume strain which is inversely related to the diameter of the particle, D. The dependence is very strong for small particle dimensions and tends to level down to an asymptotical value:

$$\Delta_{VC\,\lim} = \frac{3G_r F(\lambda_f)}{2K_r}$$

for larger particle sizes. This threshold value is related to the ratio between the shear and the bulk modulus of the rubber and indicates that particles with the lowest shear moduli will cavitate first. It is to be noted the similarity with Gent's result (4):

$$\Delta_{Gent} = \frac{5G_r}{2K_r}$$

for the critical volume strain for cavitation in a large block of rubber. The difference with Gent's expression probably arises from the fact that, in his work, the elongation of the rubber near the hole can reach any value, while in the present approach a maximum value λ_f is considered, as an intrinsic limit of the deformability of the rubber. It is worth noting that the limit of $3/2*F(\lambda_f)$ for extremely large λ_f is 2.45, very close to Gent's factor $5/2$.

For rubbers typically blended with polymers, Δ_{VClim} becomes negligible for diameters below 400 nm and eq 3 can be simplified as:

$$\Delta_{VC} \cong 4\left(\frac{4\Gamma}{3K_r D}\right)^{3/4} \tag{4}$$

According to eq 4, it is also possible to predict that, for polymer blends with a distribution of particle sizes, cavitation will begin in the largest particles and progressively will affect the smaller ones. Since for most polymers, the yield or fracture stresses are in the range 30-70 MPa, eq 4 enables to predict that if the particle size is below 0.8-0.3 μm, the polymer will undergo yielding or fracture without cavitation. For rubber toughened nylon tested in impact at 23°C, the critical particle size is around 0.2

μm. For smaller particles, the cavitation stress will be higher than the yield or fracture stress.

The experimental data in Figure 1 are taken from the works of Dompas *et al.* (*5*) and Schwier *et al.* (*6*). The theoretical line was calculated by using eq 4 with the following parameters: $G = 0.4$ MPa, $F(\lambda_f) = 1$, $K_r = 2$ GPa. The surface energy Γ has been considered as an adjustable parameter and the best fit value has been found to be 75 mJ/m^2, almost coincident to that suggested by Dompas *et al.* (*5*) to account for the additional contribution due to breaking of chemical bonds during cavitation. This approach was also adopted by Kramer in his analysis of crazing in glassy polymers (*7*): bond rupture was estimated to increase Γ from 40 to 87 mJ/m^2 in polystyrene, which is uncrosslinked. In rubbers with high degrees of crosslinking, bond rupture could raise Γ substantially.

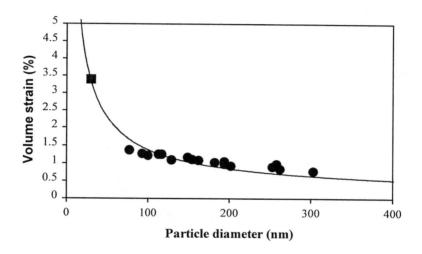

Figure 1. Critical volume strain versus particle diameter: (●) data from Dompas et al. (ref 5); (■) data from Schwier et al. (ref. 6). The theoretical line is obtained by eq 3.

Another study of cavitation in rubber toughened polymers has recently appeared (*8*). Although the expression for the stored strain energy differs from eq 2, this work confirms that the critical stress for cavitation depends on particle size, at least for particles with a diameter below 10 μm. Above this value the contribution of the first term in eq 3 becomes negligible compared to the Gent's term, and the cavitation stress becomes almost independent of particle size.

Argon *et al.* (*9*) pointed out the role of thermal mismatch in rubber toughened polymers. For polybutadiene particles in polystyrene, the hydrostatic stress resulting from the mismatch of the coefficients of expansion of the matrix and of the rubber can reach 28 MPa (*9*). Eq 4 suggests that, for a typical rubbery phase and this level of thermal stress, all particles with a diameter above 0.1 μm should undergo cavitation during the cooling stage following the moulding of the material. Recent work (*10*) has shown that the release of energy is a necessary but not sufficient condition for cavitation, and that a small energy barrier at small void sizes exerts a controlling influence upon the cavitation process. This barrier arises from the non-monotonic dependence of *W* from the void size *d*, and its height decreases as volume strain applied to the particle increases. This barrier to cavitation is therefore progressively reduced as the temperature is lowered, since the yield stress of the matrix increases. Moreover, it is likely that only some of the particles need to cavitate since the consequent release of elastic energy from the matrix may mitigate the effect of the thermal mismatch.

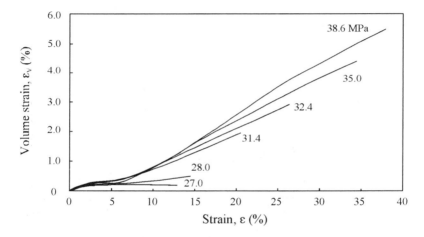

Figure 2. Volume strain versus longitudinal strain during creep tests for rubber toughened nylon66/20EPR-g-MA. (Reproduced with permission from reference 11. Copyright 1999 Kluwer Academic Publishers.)

The prediction of the model outlined above can explain a number of experimental facts. As an example, in Figure 2 is shown the volume change of a rubber toughened nylon 66 (rubber volume fraction $\Phi_R = 0.2$) in a creep test. Below 27.0 MPa there is only an increase in volume in the elastic range, while in the plastic region the volume is constant. As the creep stress is raised above this value the volume of the sample increases also in the plastic region, due to the nucleation and growth of microvoids within the rubber particles. The volume strain[*] of the sample ε_V is proportional to the longitudinal strain ε with a rate that increases with the applied stress (Figure 3). In fact, the slope $d\varepsilon_V/d\varepsilon$ increases rapidly at stresses above 27 MPa and tends to level out at a value of about 0.2. The cavitation model can explain this experimental observation taking into account the particle size distribution of the blend (Figure 4). By using the relation between the cavitation stress and the diameter of the particles (eq 4) it is possible to calculate the number of particles that are able to cavitate at a given level of stress.

Figure 3 shows how the curve representing the number of cavitated particles as a function of the applied stress superimposes completely over the curve representing the dependence of the slope $d\varepsilon_V/d\varepsilon$ on the applied stress. The asymptotical value 0.2 corresponds to the volume fraction Φ_R of the rubbery phase in this toughened nylon and is reached when all the particles have cavitated. Figure 3 suggests the following relationship:

$$\frac{d\varepsilon_V}{d\varepsilon} = n_C \Phi_R \qquad (5)$$

where n_C is the fraction of particles which have cavitated. Considering a simple model of the deformation of a cavitated rubber particle it is also possible to derive the previous expression. In fact, the small spherical void formed within the particle tends to expand as the material is further stretched. At low plastic strain it tends to assume an ellipsoidal shape and later, with increasing stretching, a "sausage-like" shape as shown by Boyce and coworkers (12), due to the anisotropic plastic behavior of polymers. The volume change of each cavitated particle, due to the plastic flow of the matrix, can be estimated by the expression:

$$\Delta_V \approx \frac{\frac{\pi}{6} D^2 \Delta D}{\frac{\pi}{6} D^3} = \varepsilon$$

where ΔD is the deformation of the particle along the stretching direction. Taking into account the volume fraction of the particles and the number of those which have cavitated, eq 5 follows immediately.

Upon integration of equation 5 we get:

$$\begin{cases} \varepsilon_V = n_C \Phi_R (\varepsilon - \varepsilon_C) & \text{for } \varepsilon > \varepsilon_C \\ \varepsilon_V = 0 & \text{for } \varepsilon < \varepsilon_C \end{cases} \qquad (6)$$

where ε_C is the longitudinal strain corresponding to the nucleation of the first voids in a tensile test. Figure 5 shows the true stress-strain curve for a water equilibrated toughened nylon 6-6 containing 20% by volume of EPR-g-MA rubber. The material shows a certain degree of strain hardening with no stress maximum. In the same figure,

---

[*] Here the symbol Δ_V is used for the volume strain of a single particle while the symbol ε_V refers to the volume strain of a whole specimen.

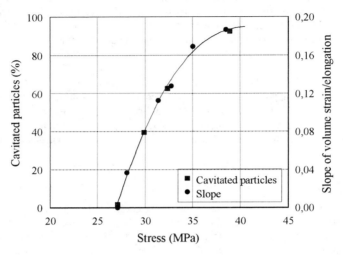

Figure 3. Slope of ε_v - ε curves in creep test and number of cavitated particles versus applied stress for rubber toughened nylon 66/20EPR-g-MA.

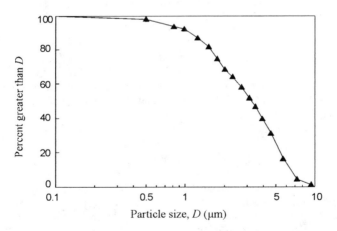

Figure 4. Particle size distribution for rubber toughened nylon66/20EPR-g-MA. (Reproduced with permission from reference 11. Copyright 1999 Kluwer Academic Publishers.)

the volume strain, ε_V, is reported as a function of the longitudinal strain. Eq 6 seems to apply at limited values of strain, the deviation at higher strains is probably associated with the strain hardening of the material which causes a rise in the stress level and, consequently in the number of cavitated particles. A critical strain of about 6-7% (corresponding to a stress of ~27MPa) and a slope $d\varepsilon_V/d\varepsilon = 0.18$ could be evaluated by drawing the tangent of the ε_V - ε curve when this starts to increase rapidly.

These values should not be regarded as material constants since they are dependent upon the interactions between the cavitation resistance of the rubber and the yielding and post-yielding behaviour of the matrix. When the temperature is lowered or at high strain rates, the yield stress of the matrix will be higher and different values for ε_C and $d\varepsilon_V/d\varepsilon$ are to be expected. As an example, Figure 6 shows the σ-ε behavior and the ε_V - ε curve for the same material shown in Figure 5, with the only difference that the material has been maintained under dry conditions after molding, therefore it was not allowed to absorb water which is known to have a plasticization effect in nylons. For this reason, the matrix shows a substantial increase in its yield stress while the rubbery phase, whose behavior is not influenced by the water content, has not changed its cavitation resistance. A decrease in the critical strain for cavitation and a parallel increase in slope of the ε_V - ε curve are to be expected, in relation with the higher fraction of particles which undergo cavitation as a consequence of the increased level of stresses in the dried toughened nylon. This is confirmed from the data in Figure 6, which show a value of ~4% for ε_C and a slope $d\varepsilon_V/d\varepsilon$ of about 0.20.

Dilatational Yielding

When a rubber toughened material is subjected to an external load, during the earlier stages of deformation, the hydrostatic component of the stress in the material starts to build-up and at a certain point, when the critical conditions discussed in the preceding section are met, the biggest particles will start to cavitate. In this initial stage, voids will appear randomly but their presence significantly affects the yielding and fracture behaviour of polymers (*13*). Lazzeri and Bucknall (*1,2*) proposed a modified version of the Gurson yield function (*14*) to account for the effects of cavitation on the yielding behaviour of rubber toughened polymers:

$$\sigma_e(\phi_R, f) = \sigma_o(\phi_R, 0)\sqrt{\left(1 - \frac{\mu\sigma_m}{\sigma_o(\phi_R, 0)}\right)^2 - 2fq_1\cosh\left(\frac{3q_2\sigma_m}{2\sigma_o(\phi_R, 0)}\right) + q_3 f^2} \qquad (7)$$

where Φ_R is the rubber volume fraction, σ_e (Φ_R, f) is the effective (von Mises) yield stress and σ_o $(\Phi_R, f=0)$ is the matrix yield stress when the mean normal stress σ_m and the void content f are both zero, while μ is the pressure coefficient of yielding. Factors q_1, q_2 and q_3 were introduced originally by Tvergaard (*15*) to improve the fit between Gurson's predictions and finite element analysis (FE) studies on metals. In its original form, factors q_1, q_2 and q_3 in eq 7 were all taken as equal to 1.

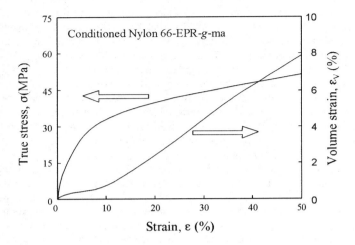

Figure 5. True stress and volume strain versus elongation for conditioned toughened nylon 66/20EPR-g-MA. (Reproduced with permission from reference 11. Copyright 1999 Kluwer Academic Publishers.)

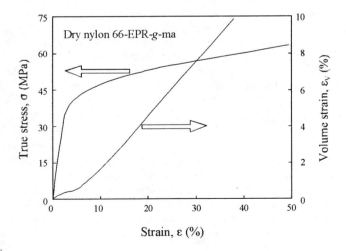

Figure 6. True stress and volume strain versus elongation for dry toughened nylon 66/20EPR-g-MA. (Reproduced with permission from reference 11. Copyright 1999 Kluwer Academic Publishers.)

Jeong and Pan (*16*) and Steenbrink and van der Giessen (*17*) also proposed other versions of the modified Gurson equation with more complex hyperbolic cosine arguments to account for the dilatancy of pressure-sensitive matrices and for elastic effects in the evolution of void volume fraction. It is to be noted that, at low levels of triaxiality, $\cosh(x) \cong 1 + x^2/2$ and all these expressions differ by irrelevant amounts. In the equation proposed by Steenbrink and van der Giessen (*17*) a major source of deviation from eq 7 is caused by the absence of a pressure dependent term.

Recent experimental work has shown that eq 7 tends to overestimate the yield stress of rubber toughened epoxies subjected to biaxial stress states (*18*). This is probably due to the approximations made by Gurson in his original work, where he considered low void volume fractions.

A 3D finite element analysis has been carried out using ANSYS (*19*) in order to obtain suitable values for parameters q_1, q_2 and q_3 and to improve the agreement with experimental results. A simple cubic array of particles has been considered and by using the symmetry of the geometry and of the applied loads only one-sixteenth of the cube has been analyzed (*16, 20*).

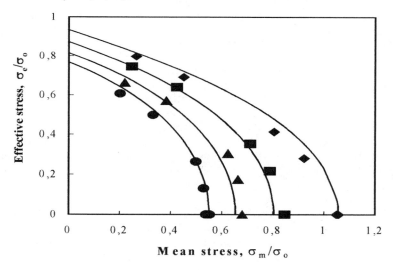

Figure 7. Yield stress at different triaxialities, according to 3D finite elements calculations for a rubber toughened polymer at: ($\blacklozenge$) $f = 5\%$, ($\blacksquare$) $f = 10\%$, ($\blacktriangle$) $f = 15\%$, ($\bullet$) $f=20\%$. *The continuous lines are calculated by using eq 7 with* $q_1=1.375$, $q_2=1.01$ *and* $q_3=2.988$.

Suitable constraints have been chosen to ensure that the faces of the cell remain planar and mutually orthogonal during the loading process. The polymer matrix has been modeled as an elastic-perfectly plastic Drucker-Prager material with a Poisson ratio ν of 0.4, a yield stress of 70 MPa (in uniaxial tensile tests), a Young's modulus E of 3 GPa and a pressure sensitivity factor μ of 0.39. The finite element model consists of 1521 SOLID92 3-D 10-node tetrahedral elements with 2728 nodes. SOLID92 has a quadratic displacement behavior with large strain capabilities. Numerical computations were carried out at five different values of the triaxiality ratio $T = \sigma_m/\sigma_e$ and at four different void fractions (f = 5%, 10%, 15% and 20%). The macroscopic effective stress σ_{ey} at yielding is taken as the maximum of the macroscopic stress-strain curve. The macroscopic hydrostatic stress σ_m is that corresponding to the same macroscopic strain ε_e which is associated to the maximum stress in the macroscopic stress-strain curve. These computed values, after being normalized by σ_o are shown in Figure 7. The best fit of the FE computation data suggests the values q_1 = 1.375 and q_2 = 1.010, q_3=2.988, which are close to those proposed by Tvergaard (*15*).

Comparison with Experimental Results

Very few data are available in the literature on the yielding behavior of rubber toughened polymers tested under a multiaxial state of stress. A pioneering paper by Sultan and McGarry (*21*) examined the yield behavior of two rubber toughened epoxies and of the corresponding unmodified epoxy under biaxial stress system and provides the basis for a first comparison of the modified Gurson equation with experimental data.

The rubber modifier was a carboxyl terminated butadiene-acrylonitrile copolymer (CTBN) and the volume fraction was about 10% (*21*). By changing the butadiene/acrylonitrile ratio, and thus the solubility parameter of the elastomer, these authors were able to obtain two rubber toughened epoxies with a considerably different particle size distribution. In particular, the rubber containing 25% acrylonitrile, had an average diameter below 100 nm (defined "small" particles in the following discussion) while the CTBN with an acrylonitrile content of 18% showed particles in the range 0.5-1.5μm ("large" particles). Sultan and McGarry (*21*) found that the unmodified epoxy followed the modified von Mises criterion with $\sigma_o(0, 0)$ = 76 MPa and with a pressure dependence parameter μ=0.371. The yield curve for a pure epoxy calculated from the data reported by Sultan and McGarry is shown in Figure 8 (curve 1). The epoxy toughened with "small" particles showed a similar behavior with a lower σ_o = 63 MPa, due the stress concentration of the rubber particles, but the same value of μ (curve 2 in Figure 8). Electron microscopy showed that the material with "small" particles did not show evidence of cavitation at these levels of stresses (*21*) since the particle diameter was below 200 nm, as predicted by the cavitation model discussed above. For f=0, eq 7 gives a linear dependence of the effective (von Mises) yield stress σ_e at yielding from the hydrostatic stress σ_m. The line corresponding to the toughened epoxy is only shifted downwards, for to the stress concentration (SC) in the matrix due to the presence the rubber, but parallel to that of unmodified epoxy, because the SC factor is independent of the level of triaxiality. When the epoxy modified with "large" particles was tested at negative values of the hydrostatic stress σ_m, the experimental points fell on the same line as that for the material with "small" particles (*21*). This can clearly be explained by the fact that voids can not be nucleated under negative hydrostatic stresses.

For positive hydrostatic stresses however, Sultan and McGarry (*21*) could not fit their data with the same line used for "small" particles and proposed a new line with μ=0.445 (curve 3, Figure 8). Electron microscopy showed clear evidence of rubber particle cavitation in this toughened epoxy (*21*). The individual yield points for this system with "large" particles obtained by these authors are also reported in Figure 8, together with the yield curve (curve 4) calculated by means of eq 7 with f = 0.1 and the same σ_o value obtained for "small" particles. Despite the relative large scatter of the experimental data, it can be observed how the non-linear modified Gurson equation provides a better fit than the linear expression proposed by Sultan and McGarry (*21*).

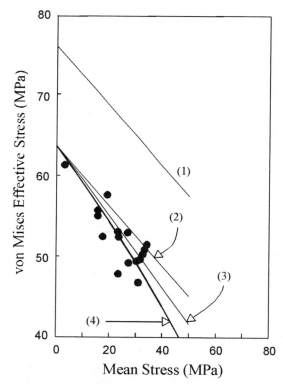

Figure 8. Yield curves for unmodified and rubber toughened epoxy resins in the σ_e-σ_m plane (see text for their numbers). Experimental points from Sultan and McGarry (ref. 21).

These authors (21) also reported that the activation volume of the epoxy modified with big particles was 1.38 times larger than that measured for the system with small particles. It is now possible to assess that this is also a consequence of rubber particle cavitation and this behavior will be discussed in the next section.

Kinetics of Yielding

The effects of voids on the apparent activation volume can be calculated by using the modified Gurson yield function. By substituting the values $\sigma_m = \sigma_y/3$, $\sigma_e = \sigma_y$ as in a tensile test, and the approximation $\cosh(3q_2\sigma_m/2\sigma_0) = 1 + 1/2(3q_2\sigma_m/2\sigma_0)^2$, from eq 7 we find:

$$\sigma_y = \sigma_o \frac{12\mu - \sqrt{144\mu^2 - 36(4\mu^2 - 9fq_1q_2^2 - 36)(1 - 2fq_1 + f^2q_3)}}{4\mu^2 - 9fq_1q_2^2 - 36} \tag{8}$$

where σ_y is the yield stress in a tensile test, and σ_o (Φ_R, $f=0$) follows the Eyring kinetics:

$$\sigma_o = \frac{RT}{\gamma V^*}\left[\frac{\Delta H}{RT} + \log\left(\frac{\dot{\varepsilon}_y}{\dot{\varepsilon}_o}\right)\right] \tag{9}$$

In the last equation, ΔH is the activation energy, V^* is the activation volume of the matrix, γ is the stress concentration factor due to the presence of the rubbery particles, R is the gas constant, $\dot{\varepsilon}_y$ is the strain rate at yielding and $\dot{\varepsilon}_o$ is a reference strain rate.

The apparent activation volume is normally calculated from the slope of the yield stress σ_y plot versus $\log \dot{\varepsilon}_y$:

$$\frac{RT}{\gamma V_{app}} = \frac{\partial \sigma_y}{\partial \log \dot{\varepsilon}_y} = \frac{\partial \sigma_y}{\partial \sigma_o}\frac{\partial \sigma_o}{\partial \log \dot{\varepsilon}_y} = \frac{\partial \sigma_y}{\partial \sigma_o}\frac{RT}{\gamma V^*} \tag{10}$$

Since the stress concentration factor for a rubber particle is approximately the same for a void, from eq 10 the ratio of the apparent activation volume for a polymer where the particles have cavitated, $V_{app}(f)$, to the apparent activation volume for a polymer where the rubbery phase is not cavitated, $V_{app}(0)$, can be estimated as:

$$\frac{V_{app}(f)}{V_{app}(f=0)} = \frac{3}{3 + \mu} \frac{4\mu^2 - 9fq_1q_2^2 - 36}{12\mu - \sqrt{144\mu^2 - 36(4\mu^2 - 9fq_1q_2^2)(1 - 2fq_1 + f^2q_3)}} \tag{11}$$

A plot of this ratio as a function of the volume fraction of voids f, shows an approximately linear relationship with a slope 2, for $\mu = 0.37$-0.39. This means that for a volume fraction of 0.1, the ratio is predicted to be about 1.2. The value of 1.38 measured by Sultan and McGarry (21) is quite higher and is probably due to experimental error due to the technique used to measure the activation volume. In their work, they performed tensile tests over 5 decades of strain rates, and the calculation of the slope by using such a limited number of decades can lead to substantial scatter in the data. For example they also reported (21) that the activation volume of the epoxy with the small particles was 14% smaller than the unmodified resin. This is unlikely, because of the stress concentration associated with the rubber particles and the slight plasticizing effect of the CTBN rubber, due to the reaction of the carboxyl groups with the epoxy matrix, which leads to some degree of copolymerization.

Lazzeri and Giuliani (22) studied a similar system to that used by Sultan and McGarry (21), but used a different technique to measure the activation volume, in order to avoid the difficulties mentioned above.

A Shell Epon 828 epoxy resin, crosslinked with 5phr piperidine, was considered. The rubber used was a BF Goodrich CTBN 1300X8, with an acrylonitrile content of 18% and three blends were prepared with a rubber content of 5, 10 and 15 phr. After blending the epoxy and the rubber at about 60°C, the hardener was added and the system was cured at 100°C for 14 hours. A post-cure cycle at 130°C for 4 hours followed. The particle size was about 1-2μm and a SEM analysis showed that the particles cavitate when tested in tension before reaching the yield stress.

The activation volume was measured by the stress relaxation technique (22) both in tension and compression for all the blends and for the unmodified epoxy. Since cavitation is not possible in compression, for each composition the corresponding value of the activation volume will be considered as V_{app} (0), while the value of activation volume in tension will be V_{app} (f), where the volume fraction of voids f will be assumed equal to the volume fraction of the rubber in the blend, since the particles are now cavitated.

Table I reports the activation volume data for the CTBN/epoxy system. The ratio V_{app} (f)/ V_{app} (0) is calculated both from the experimental results and from eq 10.

Table I. Apparent Activation Volume for Epoxy/CTBN Blends

Formulation (phr)	$V_{app}(Å^3)$ Compression	$V_{app}(Å^3)$ Tension	Ratio experimental	Ratio Calculated
Neat Epoxy	1078(*)	1089(*)	1.01	1
Epon 828/5 CTBN	1197	1349	1.13	1.09
Epon 828/10 CTBN	1316	1596	1.21	1.19
Epon 828/15 CTBN	1435	1863	1.30	1.30

(*) Extrapolated value.

As shown in Table I, the difference between the calculated and experimental ratio is below 3%. This result clearly shows that the presence of voids significantly affects the rate of yielding as indicated by the increase in apparent activation volume. Since the method used for the activation energy requires the previous calculation of the activation volume, it is quite clear that also the apparent ΔH will be lower for a polymer where the rubber particles cavitate prior to matrix yielding, as measured by Sultan and McGarry (21).

Table II. Apparent Activation Parameters for Nylon 66/EPR-*g*-MA

Formulation (wt%)	V_{app} ($\mathring{A}^3$) Tension	ΔH (kJ/mol) Tension
Neat Nylon 66	1130	205(*)
Nylon 66/5EPR-*g*-MA	1620	227
Nylon 66/10EPR-*g*-MA	1870	218
Nylon 66/15EPR-*g*-MA	2030	201
Nylon 66/20EPR-*g*-MA	2030	184

(*) For pure nucleated nylon 66 a value of 230 kJ/mol was obtained.

In Table II apparent activation volume and apparent activation energy data measured in a series of tensile tests over a range of strain rates and temperatures for rubber toughened nylons at different compositions are presented. Due to the cavitation process, the apparent activation energy shows the expected decrease with increasing rubber content, although the neat nylon presents a lower value compared to the blends containing 5 and 10% of EPR-*g*-MA. This discrepancy might be explained with the fact that neat nylon has a spherulitic structure, while the presence of the rubbery phase significantly affects the crystallization behavior, leading to preferential crystal orientation at the interface, although the degree of crystallinity of the nylon matrix was practically unaffected (*11*). To evaluate the role of crystallization conditions, the activation energy of the same nylon matrix, molded in presence of a commercial nucleating agent, was also measured and a value of 230 kJ/mol was obtained for this material. It is known, in fact, that the spherulite size has a marked influence on the properties of nylons. A smaller spherulite size leads to higher modulus and yield stress, and has a negative effect on fracture toughness.

The above discussion has shown that cavitation of the rubber particles, not only influences the value of the yield stress but also the kinetics of yielding, through its effect on both V^* and ΔH. In the past this behaviour was not associated with the presence of voids in the particles and no explanation had been given for the experimental results. Since the stress concentration in the matrix around a void or a rubber particle is approximately the same, the increase in the apparent activation volume can be only explained by the fact that a void allows plastic flow of the matrix around it, more easily that an intact rubber particle. In fact, the bulk modulus of a rubber particle is very high while for a void is zero. Thus the increased local strain rate of the polymer matrix is due to the fact the material is no longer a "real" continuum on a microscopic scale. The deformation involves a macroscopic volume increase due to the growth of the voids generated inside the rubber particles, which is favored by high level of triaxiality, leading to a non-linear yield curve. In contrast to the case of a "continuum" polymer, where high levels of triaxiality favor crazing and cleavage mechanisms over shear yielding, for a "porous" polymer a high triaxiality considerably accelerates plastic flow around voids and cause a lowering of the macroscopic yield stress, resulting in a observed value of V^* and ΔH.

Plastic Zone Size and Impact Strength

As discussed above, cavitation in the rubber particles causes a lowering of the macroscopic yield stress that is more evident at higher levels of triaxiality, as near a crack tip where it leads to an enlarged plastic zone. This effect is particularly important in rubber toughening of polymers because in many systems the impact strength has been found proportional to the volume of the plastic zone (23-24).

A first order approximation of the size and shape of the plastic zone in a rubber toughened polymer can be calculated on the basis of the modified Gurson yield function (eq 7) by considering an elastic analysis. Although this approach underestimates the real size of the plastic zone since it does not take into account the stress redistribution caused by local plastic flow, and the formation of *dilatation bands* (discussed in the next section), in which the level of cavitation is well above the average, and strains are sufficiently high to cause significant departures from linear elastic behaviour, it provides a valuable insight into the behaviour of rubber-toughened plastics in notched fracture tests.

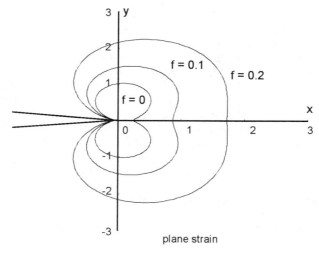

Figure 9. Approximate shape of the plastic zone at different void contents (adimensional coordinates).

Figure 9 compares the shape of the plastic zone near a crack tip for a rubber toughened polymer in condition where the particles have not cavitated ($f = 0$) with that of the same polymer at two different levels of microvoid content ($f = 0.1$ and $f = 0.2$) in plane strain. As it is evident from the figure, the size of the plastic zone increases considerably with the volume fraction of microvoids. Since, during an Izod or Charpy impact test, most of the energy is dissipated within the plastic zone, it can be anticipated that a polymer with rubber particles which are not able to cavitate will develop a small plastic zone and will show a low value of impact strength.

This is a consequence of the highly non linear dependence of the yield stress on the hydrostatic stress for a polymer with cavitated particles (eq 7). The effect of cavitation is quite limited in tensile test (low triaxiality), so that the yield strength of the polymer is not reduced excessively, while the nucleation and growth of voids has a dramatic effects on the yield stress near a notch or a crack tip (high triaxiality) leading to the formation of a large plastic zone and hence to a high value of fracture resistance. This is the main difference between rubber toughening and plasticizer toughening. In the latter case, a plasticizer is added to the matrix resulting in a reduction of yield stress at all levels of triaxiality. In other words, the yield curve (for example line 1 in Figure 8) is shifted parallely downwards, so that a high value of fracture resistance is obtained at the expense of a large reduction of the tensile resistance of the material.

Figure 9 also shows that moderate levels of cavitation reduce but do not eliminate the effects of plastic constraint on the central portion of thick notched bars.

Purcell (25) in his earlier study on the notched Izod impact behaviour of PVC-MBS blends showed a strong particle size effect in thick (6.4 mm) bars, with the impact energy falling from 1200 J/m for a blend with an average particle diameter $D = 270$ nm ("large" particles in the following discussion) to 200 J/m for a material with $D = 57$ nm ("small" particles), whereas in thin (3.2 mm) bars impact energies decreased much more slowly, from 1300 to 1100 J/m for the same materials.

These results can now be understood on the basis of the cavitation model. The thickness effect is very pronounced when particle sizes are small, and the cavitation stress is very high resulting in a limited amount of cavitation. In fact, the 6.4 mm thick specimens are in a plane strain situation, and if the rubber particles are not capable of cavitating (diameter too small), the plastic zone will be very limited and so the impact strength of the material. By contrast, the 3.2 mm bars are in a plane stress condition and the plastic zone will be large enough to ensure a relatively high level of impact strength, even in the absence of cavitation.

For the material with "large" particles, cavitation of the rubbery phase leads to the formation of a large plastic zone even for the thick samples, and the impact strength of the material when molded into thick specimens will be almost equal to that measured on thinner bars.

From these results it is clear that the transition from plane stress to plane strain conditions may cause a large reduction in toughness in rubber toughened polymers, when the particle size is too small for cavitation to occur. On the other hand, thickness effects may be negligible when the particles are large enough to cavitate before the material reaches its yield stress.

Dilatational Bands

The work of Pearson and Yee on rubber-toughened epoxy resins has shown that particle cavitation begins at an early stage in the deformation (26, 27). Initially, the process appears to occur with no particular correlation between sites ("random" cavitation). This is to be expected because of the inverse relationship between the critical mean stress at cavitation ($\sigma_c = K_r \Delta_{VC}$), and particle diameter D (eq 4). In real materials, the particle size distribution can be relatively large and the larger particles will cavitate earlier.

The nucleation and growth of voids in the particles, as shown before, substantially alter the yield behavior of a polymer, and in particular cause the transition from a linear to a non-linear yield function. It must be pointed out that the modified Gurson expression (eq 7) can only model the initial yielding of the material under conditions of "random" cavitation of the particles. Further deformation is no longer homogeneous and becomes highly localized due to the initiation of shear instabilities called *dilatational bands*. These bands can be described as planar arrays of cavitated particles where the matrix between voids is subject to high shear strains and, because of rubber particle cavitation, reduced constraints on plastic flow.

As shown in previous papers (1, 2), by considering that the only types of deformation are simple shear parallel to the plane and volume dilatation normal to it, it is possible to calculate the angle Ψ between these dilatational bands and the direction of maximum stress on the basis of the modified Gurson relation (eq 7):

$$\cos 2\Psi = -\frac{2\alpha\sigma_o}{\sigma_1 - \sigma_2} \tag{12}$$

where σ_1 and σ_2 are the principal stresses on the plane of the band and

$$\alpha = \frac{\mu}{3}(1 - \mu\frac{\sigma_m}{\sigma_o}) + \frac{f}{2}q_1q_2 sinh\left(\frac{3q_2\sigma_m}{\sigma_o}\right) \tag{13}$$

The angle of the band, Ψ, is thus a function of the void content f and of the level of triaxiality (1-2). For a typical rubber toughened polymer containing 10% by volume of particles, the angle $\Psi \sim 30°$ in a tensile test and near zero in proximity of notches and cracks (28-30).

Once the deformation process has become localized into dilatational bands, further particle cavitation will occur preferentially near band tips ("cooperative" cavitation as discussed below). The cavitation process is no longer random and eq 4 cannot accurately account for the macroscopic yield behaviour of the cavitated polymer.

Propagation and Growth of Dilatational Bands

The large elastic displacements at the tips of the propagating deformation bands accelerate both cavitation in the rubber particles and yielding in the matrix phase, with the result that most of the cavitated particles formed at this stage of deformation are associated with dilatational bands.

An interesting and detailed TEM study has recently been carried out by Hiltner and coworkers (*30*, *31*). This work has examined the nucleation and the growth of dilatational bands in polycarbonate toughened with core shell rubber particles (0.2 μm in diameter) at various locations of the whitened zone near a semicircular notch, after loading the sample at different levels of stress. This group of researchers has found, during the loading stage, that nucleation of microvoids initially occurs at about 1 mm from the root of the notch when the mean stress is higher. On increasing the applied load, small clumps of 2 to 10 cavitated particles appear. At this stage, further increases in stress level lead to the development of domains containing 1 to 4 cavitated particles in the width and 8 to 35 cavitated particles in the length, which can be considered mature dilatational bands. Most of the particles outside the bands were not cavitated, nor distorted (*30*), while those inside the bands appeared distorted due to the high level of strain of the matrix in the ligaments, that was estimated to be close to the natural draw ratio of PC, about 100% (*30*). The maximum density of cavitated domains was measured to be of around 0.67μm^{-2} (*31*). Cooperative cavitation in rubber-modified PC becomes more likely as the interparticle distance decreases (*30*).

The propagation of dilatation bands appeared to be related to a mechanism of "cooperative cavitation" due to the enhanced volume strain at the two ends of the band which causes the particles near the tip to cavitate. This process is similar to the growth of crazes by "repeated cavitation" of the hetero-phases described and modeled by Argon *et al.* (*9*).

A modification of Argon's approach has recently been proposed (*32*) to model the propagation of a dilatational band. The stress field at the tip of a dilatational band was calculated, as well as the strain energy ahead of the band. The volume strain energy density was found to be inversely related to the distance to the tip of the band. Cavitation, therefore, will occur preferentially in rubber particles located near the band, within a distance δ from the band tip (*32*):

$$\delta = \frac{1-2v}{6E} \frac{a}{K_r \Delta_{VC}^2} \sigma_{db}^2 Y(\theta) \tag{14}$$

where $Y(\theta)$ is a function of the angle θ from the band tip, a is the length of the band and:

$$\sigma_{db} = \sqrt{(\sigma - \sigma_d \cos^2 \Psi)^2 + (\tau - \frac{1}{2}\sigma_d \sin 2\Psi)^2} \tag{15}$$

is the local stress on the band, resulting from the stress σ normal to the band, the shear stress τ and the drawing stress σ_d of the matrix in the ligament between neighboring cavitated particles. In eq 14, E and v are the Young's modulus and Poisson's ratio of the rubber-matrix system, calculable from composite theory (*9*, *33*).

Eq 14 shows that the enhanced volume strain energy near a band will cause cavitation of bigger particles (low Δ_{VC}) even located at relatively large distance from the band, while small particles (high Δ_{VC}) need to be quite close to the band tip, and preferentially along its axis ($\theta = 0$), to be able to cavitate. As a consequence, dilatational bands are more clearly observed in those systems where the particle size distribution is rather narrow. On the contrary, when the particle size distribution is fairly large, cavitation will occur in the biggest particle at large distance from the band and not necessarily located along its longitudinal axis. In this case, a microscopic examination of the whitened zone of a polymer blend might not be able to identify linear arrays of voids, and the cavitation process will appear more of a random nature.

From eq 14, the rate of propagation of a dilatational band, da/dt can be evaluated as (9):

$$\frac{da}{dt} = \frac{\delta}{\lambda_d - 1} \dot{\varepsilon}$$

(16)

where λ_d is the natural draw ratio of the matrix in the ligament and $\dot{\varepsilon}$ is the shear rate, determined by the temperature and by the level of stress in the material.

In the following discussion, for polymers presenting a distribution of particle sizes, the distance δ from the band, where the strain energy is above the critical value for rubber particle cavitation, will be assumed that corresponding to the average particle diameter.

Equations 15 and 16 mean that the process is controlled by the initial phase of random cavitation. Upon increasing the stress level, more cavitation occurs in other particles close to those which have already developed voids, and small clumps of cavitated particles form. The stress field near these clumps is enhanced, and the particles within a distance δ will experience a volume strain above the critical value given by eq 3. Therefore, other particles cavitate. If the particle size distribution is rather narrow, the stress intensity, at distances greater than δ from the clumps, is still too small for the particles to cavitate. Thus, only the particles near the tip of the clumps are likely to undergo cavitation. Planar arrays (dilatation bands) containing several cavitated particles now form. While the external load is increased the number of bands increases as their length. The process will tend to accelerate since the rate of band propagation is proportional to the length a of the band. At this stage, the deformation within the plastic and cavitated zone becomes highly localized, with the matrix within the band undergoing large shear deformation, while the matrix outside the band remains in the elastic state.

As outlined above, the matrix in the ligament among neighboring cavitated particles stretches until the maximum draw ratio of the material is attained, and this accounts for the large energy dissipation associated with dilatational bands.

Conclusions

A model for particle cavitation and dilatational yielding in rubber toughened polymers has been reviewed. An expression for the critical volume deformation as a function of particle diameter has been derived and compared with experimental data. Good agreement has been found when chain scission energy is taken into account. The stress dependence of the volume change during creep and tensile tests in rubber toughened nylon 66 can be explained in terms of the inverse relationship between particle size and critical volume strain and the particle size distribution of the polymer blend.

A finite element model was carried out to improve the accuracy of the modified Gurson yield function and the results were compared with experimental data available in the literature. Good agreement was obtained with data on rubber toughened epoxy (*21*). It was also shown how the presence of voids also affects the kinetics of yielding by increasing the apparent activation volume and decreasing the apparent activation energy.

The effect of voids on the shape and on the volume of the plastic zone was calculated on the basis of an elastic analysis and by using the modified Gurson expression. The results were discussed in relation to thickness effects in impact tests.

The formation of dilatational bands has been reviewed and compared with experimental data. The angle of the dilatational bands to the principal stress axis, predicted by the model, has been confirmed by electron microscopy and light scattering studies (*29-31*).

The propagation and growth of dilatational bands has been modeled by a modification of Argon's model for the growth of crazes by "repeated cavitation" (*9*).

Acknowledgements

The authors wish to thank Prof. Ali Argon for his helpful comments on the manuscript.

References

1. Lazzeri, A. and Bucknall, C.B., *J. Mater. Sci.* **1993**, *28*, 6799.
2. Lazzeri, A. and Bucknall, C.B., *Polymer*, **1995**, *36*, 2255.
3. Bucknall, C.B., Karpodinis, A. and Zhang, X.C., *J. Mater. Sci.*, **1994**, *29*, 3377.
4. Gent, A.N. and Lindley, P.B., *Proc. Roy. Soc. London*, **1958**, *A249*, 195.
5. Dompas, D., Groeninckx, G., Isogawa, M., Hasegawa, T. and Kagodura, M., *Polymer*, **1994**, *35*, 4750.
6. Schwier, C.E., Argon, A.S. and Cohen, R.E., *Phil. Mag. A*, **1985**, *52*, 581.
7. Kramer, E.J., *Polym. Eng. Sci.*, **1984**, *24*, 761.
8. Fond, C. and Schirrer, R., *Journal de Physique IV*, Colloque C6, **1996**, *6*, (1996) 375.
9. Argon, A.S., Cohen, R.E., Gebizlioglu, O.S. and Schwier, C.E., *Advances in Polymer Science*; Springer-Verlag, Berlin, D, 1983, *Vol. 52/53*, pp. 275-334.
10. Ayre, D.S. and Bucknall, C.B., *Polymer*, **1999**, *39*, 4785.
11. Bucknall C.B. and Lazzeri, A., *J. Mater. Sci.*, in press.

12. Tzika, P., Boyce. M.C. and Parks, D.M., to be published.
13. Sue, H.-J. and Yee, A.F., *Polym. Eng. Sci.*, **1996**, *36*, 2320.
14. Gurson, A.L., *J. Engng. Mater. Technol.*, **1977**, *99*, 2.
15. Tvergaard, V., *Int. J. Fracture*, **1981**, *17*, 38.
16. Jeong, H.-J. and Pan, J., *Int. J. Solids Structures*, **1995**, *32*, 3669.
17. Steenbrink, A.C. van der Giessen, E., and Wu, P.D. , *J. Mech. Phys. Solids*, **1997**, *45*, 405.
18. Kody, R.S. and Lesser, A.J., *Polym. Composites*, in press.
19. ANSYS User's Manual, Revision 5.2, Swanson Analysis Systems, Inc., Houston, PA, USA,1995
20. Hom, C.L. and McMeeking, R.M., *J. Appl. Mech.*, **1989**, *56*, 309.
21. Sultan, J.N. and McGarry, F.J., *Polym. Eng. Sci.*, **1973**, *13*, 29.
22. Lazzeri, A., and Giuliani, D., *Proc. 10th Int. Conf. on Deformation, Yield and Fracture of Polymers*, 7-10 April 1997, Cambridge, UK, P76.
23. S. Wu, *Polymer*, **1985**, *26*, 1855.
24. Yee, A.F., Li, D.-M. and Li, X.-W., *J. Mater. Sci.*, **1993**, *28*, 6392.
25. Purcell, T.O., *Amer. Chem. Soc. Polym. Prepr.* **1972**,*13*, 699.
26. Pearson, R.A. and Yee, A.F., *J. Mater. Sci.*, **1986**, *21*, 2475.
27. Yee, A.F. and Pearson, R.A., *J. Mater. Sci.*, **1986**, *21*, 2462.
28. Sue, H.-J., *J. Mater. Sci.*, **1992**, *27*, 3098.
29. Schirrer, R., Fond, C. and Lobbrecht, A., *J. Mater. Sci.*, **1996**, *31*, 6409.
30. Cheng, C., Hiltner, A., Baer, E., Soskey and P.R. Mylonakis, S.G., *J. Appl. Polym. Sci.*, **1995**, *55*, 1691.
31. Cheng, C., Hiltner, A., Baer, E., Soskey, P.R. and Mylonakis, S.G., *J. Mater. Sci.*, **1995**, *30*, 587.
32. Lazzeri, A., *Proc. 10th Int. Conf. on Deformation, Yield and Fracture of Polymers*, 7-10 April 1997, Cambridge, UK, P75.
33. Chow, T.S., *J. Polym.Sci.-Phys.*, **1978**, *16*, 959.

Chapter 3

Mesoscopic Localized Deformations in Rubber-Toughened Blends

K. G. W. Pijnenburg and E. Van der Giessen[1]

**Delft University of Technology, Koiter Institute,
Mekelweg 2, 2628 CD Delft, The Netherlands**

This paper is concerned with a numerical investigation of the deformation patterns in an amorphous polymer-rubber blend at a mesoscopic scale, i.e. in between the rubber particles. The computations employ a material model for shear yielding that incorporates rate-dependent yield, strain softening and anisotropic hardening at large plastic deformations. Periodic cell calculations are reported that demonstrate that shear yielding in a cavitated blend proceeds by the initiation and propagation of mesoscopic shear bands. The implications of these localized deformations on the competition with crazing are also discussed.

It is a well established fact that the fracture toughness of polymers can be greatly enhanced by adding a dispersion of rubber particles. The toughening is commonly assumed to involve a number of mechanisms: crazing, cavitation and shear yielding (*1, 2*). Cavitation of the rubber particles relieves the stress triaxiality in the matrix polymer. This suppresses the likelihood of matrix crazing and promotes plastic deformation in the matrix by shear yielding. The toughening effect is generally enhanced when a region of large plastic deformation spreads out over a large volume in the material.

Toughening in blends thus involves a range of length scales. The 'macroscopic scale' is the scale at which plastic deformation takes place in the neighborhood of a propagating crack in a blend. The relevant 'microscopic' scale here is the molecular scale at which shear yielding takes place. The intermediate 'mesoscopic' scale is the size scale at which the individual rubber particles can be distinguished. This is a crucial scale for our understanding of toughening in blends since it is at this size scale that rubber cavitation, crazing and shear yielding compete with each other

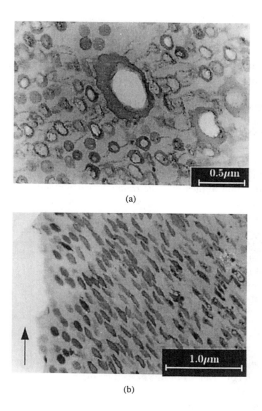

(a)

(b)

Figure 1. Deformation zones in ABS (a) at a distance of 100μm from, and (b) near the fracture surface. From (6), with kind permission from Kluwer Academic Publishers.

and determine which one(s) of them dominate(s).

Evidence of massive plastic deformation at this scale during fracture is given in the TEM micrographs in Figure 1, taken near the fracture surface in notched samples of ABS (with polybutadiene rubber particles). After cavitation, the initially spherical particles in Figure 1a are seen to develop rather bulgy shapes. Similar shapes have been observed in (3, 4, 5) in various other materials and are expected to be relevant also at some distance ahead of a crack tip. The shape of the particles near the fracture surface (Figure 1b) is quite different, however, indicating that the mesoscopic deformation processes are significantly different in this regime.

The objective of this paper is to summarize recent insight about how plastic deformation occurs in a blend at this mesoscopic scale. This insight is obtained by numerical simulation using a realistic material model for shear yielding in the amorphous glassy matrix of the blend. We shall show how plastic deformation proceeds by the propagation of shear bands in between the rubber particles. Finally, we briefly investigate the consequences of the observed localized plastic deformation on the competition with crazing in the matrix.

Material Model

We model the polymer blend as a two-dimensional material (i.e. a material with infinitely long cylindrical particles under plane strain conditions perpendicular to the long axis). To reduce the computational effort further, we take the particles to be arranged in a regular square array. Therefore we can restrict our attention to a unit cell as shown in Figure 2. The relevant parameters are the particle radius a and half of the particle spacing, b. The ratio of a and b is related to the volume (or area) fraction of particles, f, in the following way: $f = (\pi/4)(a/b)^2$.

It is commonly accepted that toughening in amorphous blends must involve the cavitation of the rubber particles or debonding of the particles from the matrix. Since the shear modulus of the rubber commonly used in blends is much lower than the yield stress in the matrix, the presence of already cavitated particles affects the subsequent deformations in the matrix only slightly. A detailed study of this in (7) has revealed that a cavitated rubber particle is mechanically equivalent to a void up to relatively large ratios between rubber modulus and matrix yield strength. Therefore, we can replace the rubber particles by traction-free voids.

The applied loading consists of macroscopic strain rates $\dot{E}_{11}$ and $\dot{E}_{22}$ in the x_1 and x_2 direction, respectively, as well as a shear rate $\dot{E}_{12}$. Therefore the macroscopic velocity gradient L_{ij} reads

$$L_{ij} = \begin{pmatrix} \dot{E}_{11} & \dot{E}_{12} \\ 0 & \dot{E}_{22} \end{pmatrix} . \tag{1}$$

These macroscopic loading conditions are applied through periodic boundary conditions for our unit cell. Details of this procedure may be found in (8).

The corresponding macroscopic stress components are obtained by averaging of the stresses at the mesoscale, σ_{ij}, as

$$\Sigma_1 = \frac{1}{2b} \int_{-b}^{b} \sigma_{11}(\pm b, x_2) dx_2 , \tag{2a}$$

$$\Sigma_2 = \frac{1}{2b} \int_{-b}^{b} \sigma_{22}(x_1, \pm b) dx_1 , \tag{2b}$$

$$\bar{\tau} = \frac{1}{2b} \int_{-b}^{b} \sigma_{12}(x_1, \pm b) dx_1 . \tag{2c}$$

We can write $\pm b$ because, by virtue of equilibrium, it does not matter whether the integration is performed over the upper or lower boundary (left/right in the case of Σ_1).

The material model taken for the polymer matrix material is a viscoplastic one, based on the following relation between the local (mesoscopic) shear stress τ and the resulting shear rate $\dot{\gamma}^p$, as originally derived by Argon (9):

$$\dot{\gamma}^p = \dot{\gamma}_0 \exp \left[-\frac{As_0}{T} \left(1 - \left(\frac{\tau}{s_0} \right)^{5/6} \right) \right] . \tag{3}$$

Here, $\dot{\gamma}_0$ is an pre-exponential factor, A is a material parameter related to the activation volume, T is the absolute temperature and s_0 is the athermal shear strength.

To include the effect of pressure and strain softening, we use $s + \alpha p$ instead of s_0, where p is the pressure and α is a pressure dependence coefficient. Softening upon yield is incorporated by letting s evolve with plastic straining from the initial value s_0 to a steady-state value s_{ss}, via

$$\dot{s} = h(1 - s/s_{ss})\dot{\gamma}^{\mathrm{P}}. \tag{4}$$

The rate of softening is governed by the material parameter h.

The one-dimensional flow rule (equation 3) is adopted in a fully three-dimensional constitutive model by letting the driving shear stress τ be determined from

$$\tau = \sqrt{\frac{1}{2}\bar{\sigma}'_{ij}\bar{\sigma}'_{ij}}, \quad \bar{\sigma}_{ij} = \sigma_{ij} - b_{ij}, \quad \bar{\sigma}'_{ij} = \bar{\sigma}_{ij} - \frac{1}{3}\bar{\sigma}_{kk}\delta_{ij}, \tag{5}$$

where b_{ij} is the back stress ($i, j \in 1, 2, 3$ and δ_{ij} is the Kronecker delta). See reference (10) for further details. The back stress accounts for the progressive strain

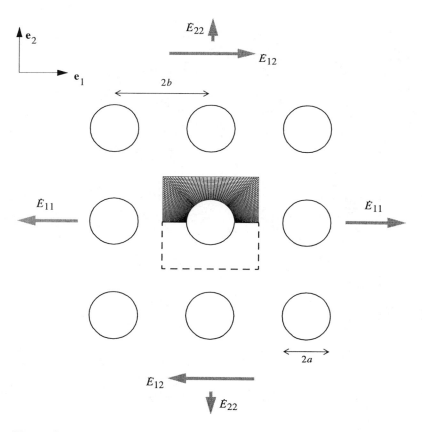

Figure 2. Sketch of the model, showing the unit cell containing a single particle/void and the finite element mesh.

hardening in amorphous polymers at continued plastic deformation. Physically, it is an internal stress associated with the stretching of the entanglement network upon continued plastic deformation. By exploiting the analogy with rubber elasticity, the back stress is modeled using non-Gaussian network theory. With reference to (7) for details it suffices to note here that the resulting constitutive description involves two material constants: the average number of links between entanglements, N, and the rubber hardening modulus C^R.

For reference purposes, the stress-strain curve of a homogeneous (i.e. un-voided) piece of matrix material under macroscopic simple shear ($\dot{E}_{11} = \dot{E}_{22} = 0$) is shown in Figure 3. Here, as well as in the sequel, we have used the effective shear strain defined by

$$\Gamma = \sqrt{E_{11}^2 + E_{22}^2 + \frac{1}{2}E_{12}^2} \tag{6}$$

as the scalar measure for the macroscopic strains. The material parameters used in the calculations are taken to be representative of SAN and are: $E/s_0 = 12.6$, $\nu = 0.38$, $s_0 = 120\,\text{MPa}$, $\dot{\gamma}_0 = 1.06 \times 10^8\,\text{s}^{-1}$, $s_{ss}/s_0 = 0.79$, $As_0/T = 52.2$, $h/s_0 = 12.6$, $\alpha = 0.25$, $N = 12.0$ and $C^R/s_0 = 0.033$ Clearly visible in Figure 3 is the drop in stress after yield which is typical for amorphous glassy polymers. This softening will cause highly localized deformations in the model blend material, as will be seen in the next Section.

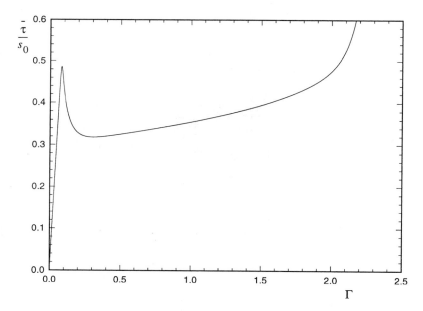

Figure 3. Shear stress ($\bar{\tau}$) response to simple shear, i.e. $\Gamma = E_{12}/\sqrt{2}$, for homogeneous SAN at a strain rate of $\dot{\Gamma} = 10^{-2}\,s^{-1}$.

Mesoscopic Shear Banding

The deformation patterns in the model blend, using the above material model for the matrix, are computed using a finite element technique with due account of large strain effects. In all calculations in this study we used a value for a/b of 0.5, corresponding to a volume fraction of $f = 20\%$.

As expected, the intrinsic softening leads to highly localized deformations in between the particles/voids. This can most easily be seen in contour plots of the plastic shear rate $\dot{\gamma}^p$. These show the locations where most plastic activity is taking place at the particular instant shown. Plastic strains are developing in narrow zones: the mesoscopic shear bands. Depending on the type of macroscopic mode of deformation (predominant tension, predominant shearing or a combination of both), different types and directions of shear band develop.

Under shear, as shown in Figure 4, the shear bands form in the ligament between the voids, parallel to the shearing direction. Once the ligaments between adjacent particles/voids are completely crossed by the shear band, large deformations on a macroscopic scale are possible. When the shear bands connect the voids, a decrease in overall stress, or macroscopic softening, is observed. This is shown in Figure 5, where the squares mark the stages of deformation at which the snapshots of Figure 4 have been made. The contour plots also show the propagation of the shear bands, once the yielded material locks due to progressive strain hardening. The bands propagate outwards from the center, thereby giving the void its particular S-like shape. This shape and its relevance have been discussed in more detail in (11). It is noted that although the material inside shear bands hardens significantly, macroscopic hardening in the overall behavior is not observed in Figure 5 because of continued shear band propagation. Macroscopic hardening under shear only occurs at very much larger strains.

Figures 6 to 8 show the development of the shear bands under different loading conditions. Just as in Figure 4, the plots were made before, at and after yield and in the stage of propagating shear bands. Besides many similarities between the figures there are also some noteworthy differences.

Uniaxial tension for the volume fraction considered here gives rise to shear bands that start near the equator of the voids, and are oriented at an angle of 30° to 45° to the tensile direction. These so-called 'dog-ear' shear bands have been shown in detail in (12). Biaxial tension, shown in Figure 6, gives a pattern of plastic flow with two families of such shear bands corresponding to tension in the x_1 and x_2 direction. Although these bands are confined to the neighborhood of the void at small strains, they are seen to link up and expand to span the entire ligament between voids upon continued deformation.

Simultaneous shear and tension/compression modes of deformation result in shear bands that start at somewhat different angles and locations along the void surface. This is shown, for example, in Figure 7 for a combination of shear and biaxial tension. However, the bands will still extend from void to void at some stage during the deformation. Again this results in macroscopic softening behaviour and the propagation of shear bands with ongoing deformation.

Another interesting situation arises if tension and compression of the same magnitude are combined (Figure 8) so as to give a state of pure shear at an angle of 45° to the (simple) shear direction considered earlier in Figure 4. Geometrically, the main difference is that in this direction the size of the ligament (the distance

42

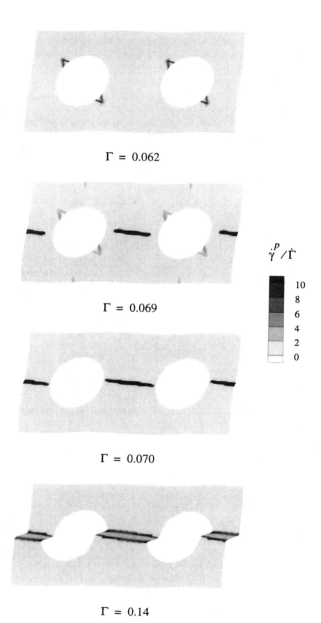

Figure 4. Distribution of instantaneous plastic shear rate $\dot{\gamma}^p$, normalized by the macroscopic shear rate, at four stages of deformation. Applied loading: simple shear at $\dot{E}_{12} = 10^{-2} \, s^{-1}$.

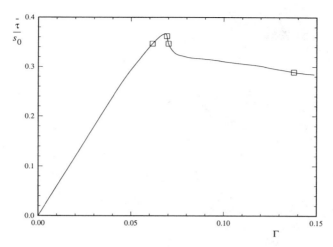

Figure 5. Average shear stress ($\bar{\tau}$) response to simple shear $\dot{E}_{12}$ for voided SAN with $a/b = 0.5$. The symbols correspond to the snapshots in Figure 4.

between two voids) is $\sqrt{2}$ times as large, since now the ligament runs diagonally through our unit cell. The evolution of the shear bands is comparable to that under simple shear (Figure 4), although due to the higher degree of symmetry there are now twice as many shear bands. They still start in the center of the ligament and propagate outward once the entire ligament is crossed, but the stress level at which yield occurs is higher. The explanation for the latter essentially lies in the larger ligament width: there is more material to resist the shear stress. We have compared the pure shear results to calculations under simple shear with the same ligament to void size ratio (i.e. a $\sqrt{2}$ times smaller volume fraction) and we found that now the macroscopic yield points are approximately the same. These findings provide evidence that the macroscopic yield behaviour of the material with a square array of particles is not isotropic. Yield is not only governed by the volume fraction but also by the void spacing in the direction where shear bands tend to develop. In general it can be said that a smaller ligament is more susceptible to macroscopic shear yielding than larger ligaments. These matters are further explored in (8).

The shapes of the voids that develop under different macroscopic modes of deformation evidently inherit from the symmetry of the morphology of the material, so that it is difficult to make direct comparison with experimental results. However, we do notice that under predominant tension, Figure 6, the voids develop into elongated shapes with distinct bulges (12). This is akin to the types of deformed cavitated particles seen in, for example, Figure 1a. On the other hand, deformation modes involving significant simple shear, such as shown in Figures 4 and 7, tend to favor the development of distinctly S-shaped voids. This is similar to the particle shapes seen close to the fracture surface in Figure 1b. We have argued elsewhere (11) that this seems to indicate that a significant macroscopic shear component may be involved in the fracture process of polymer blends as a consequence of the strain softening behavior of the matrix materials.

44

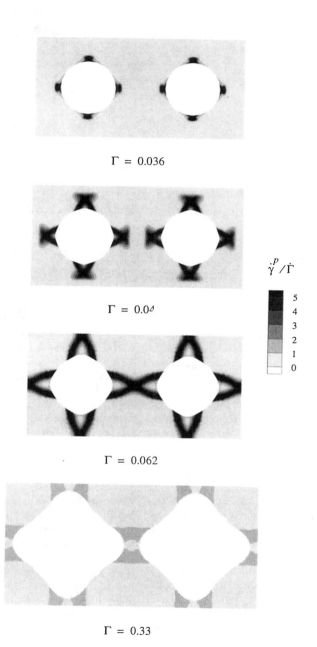

Figure 6. Distribution of instantaneous plastic shear rate $\dot{\gamma}^p$, normalized with macroscopic shear, at four stages of deformation. Applied loading: biaxial tension with $\dot{E}_{11} = \dot{E}_{22} = 10^{-2} \ s^{-1}$.

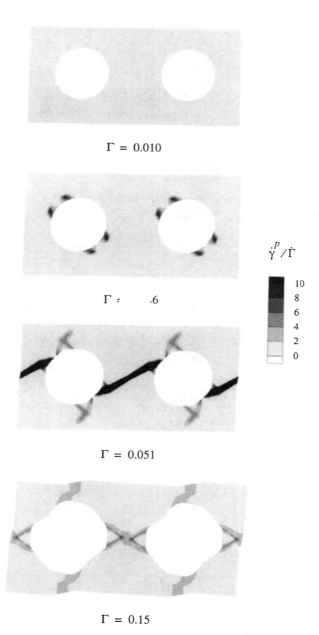

$\Gamma = 0.010$

$\Gamma = .6$

$\dot{\gamma}^p / \dot{\Gamma}$

$\Gamma = 0.051$

$\Gamma = 0.15$

Figure 7. Distribution of instantaneous plastic shear rate $\dot{\gamma}^p$, normalized with macroscopic shear, at four stages of deformation. Applied loading: simple shear in combination with biaxial tension with $\dot{E}_{11} = \dot{E}_{22} = \dot{E}_{12} = 10^{-2}\ s^{-1}$.

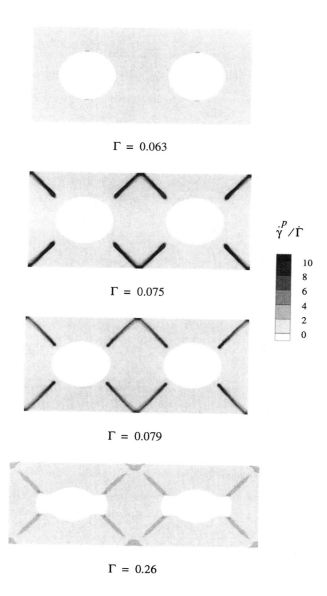

Γ = 0.063

Γ = 0.075

Γ = 0.079

Γ = 0.26

$\dot{\gamma}^p / \dot{\Gamma}$

10
8
6
4
2
0

Figure 8. Distribution of instantaneous plastic shear rate $\dot{\gamma}^p$, normalized with macroscopic shear, at four stages of deformation. Applied loading: pure shear at 45° from the x_1-axis as specified by $\dot{E}_{11} = -\dot{E}_{22} = 10^{-2} \ s^{-1}$.

Craze Initiation

In the foregoing, it has been assumed that cavitation of the rubber particles and plastic deformation by shear yielding are the only mechanisms taking place; crazing has not been considered. Toughening of a blend however depends on the competition between plastic deformation and crazing. A detailed study of this competition is still pending, but in this section we shall demonstrate how at least the initiation of crazing is dependent on the mesoscopic shear banding.

There is still much confusion about the conditions under which crazes start to develop. However, there is consensus that the hydrostatic stress $\sigma_m = (1/3)\sigma_{ij}\delta_{ij}$ plays an important role in the initiation of crazes (see e.g. (13)). In previous studies (12) we have therefore indentified regions of high hydrostatic stress in the matrix as potential locations for crazing initiation. Here, we compare such sites with the regions that satisfy an initiation criterion developed for plane stress conditions by Sternstein and Myers (14), which reads

$$|\sigma_1 - \sigma_2| \geq -A_c + \frac{B_c}{3\sigma_m}, \tag{7}$$

with the side condition that $\sigma_m > 0$. Here, σ_1 and σ_2 are the major and minor principal stress respectively, and A_c and B_c are temperature dependent constants. Noting that, under the plane strain conditions prevailing here, $|\sigma_1 - \sigma_2|$ can be approximated by $2/\sqrt{3}\sigma_e$, we can rewrite this condition as

$$S = \frac{2}{\sqrt{3}}\sigma_e - \frac{B_c}{3\sigma_m} + A_c \geq 0. \tag{8}$$

The values of the parameters A_c and B_c in equation 8 have not yet been determined experimentally for SAN, but values of $A_c/s_0 \approx 2.5$ and $B_c/s_0^2 \approx 2.0$ seem to be reasonable estimates (15) at room temperature.

Figures 9 and 10 compare the two craze initiation criteria by showing contour plots of σ_m and of S in equation 8 for two different loading situations: biaxial tension and biaxial tension combined with shear. Evidently, the regions with the highest values of σ_m or S are the critical regions for craze initiation, but whether or not crazing would actually take place depends on the values of the critical hydrostatic stress or on the values of A_c and B_c. Comparing the sets of figures we see that although the distributions of σ_m and S are different, the locations of the peak values agree rather well. Hence, the hydrostatic stress distribution can give a reasonable estimate of the critical locations for craze initiation.

The regions of the highest hydrostatic stresses in Figures 9a and 10a coincide with the locations at which the shear bands (see Figures 6 and 7) intersect. As pointed out in (12) this is a direct consequence of the kinematic 'incompatibility' created at this intersection. The intensity of this incompatibility depends on the state of stress and deformation. Both states shown in Figures 9 and 10 correspond to macroscopic yield, but the peak values of σ_m and S attained under biaxial tension in Figure 9 are significantly higher than in Figure 10 where additional macoscopic shearing is taking place. This confirms the rather intuitive notion that macroscopic shear tends to lower the susceptibility to crazing.

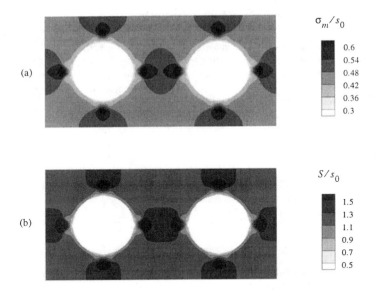

Figure 9. Distributions of (a) σ_m and (b) S, according to equation 8, at $\Gamma = 0.040$ as indicators for craze initiation. Applied loading: biaxial tension with $\dot{E}_{11} = \dot{E}_{22} = 10^{-2} \, s^{-1}$ (cf. Figure 6).

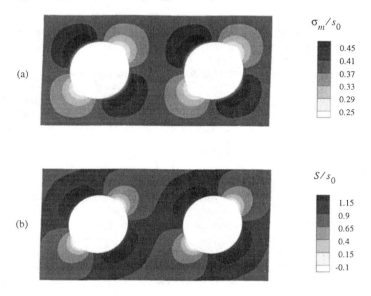

Figure 10. Distributions of (a) σ_m and (b) S, according to equation 8, at $\Gamma = 0.034$ as indicators for craze initiation. Applied loading: biaxial tension combined with shear with $\dot{E}_{11} = \dot{E}_{22} = \dot{E}_{12} = 10^{-2} \, s^{-1}$ (cf. Figure 7).

Conclusion

The computations summarized in this paper show that the mesoscopic behavior of a blend (after rubber cavitation) arises from the key characteristics of shear yielding in amorphous polymers, viz. the softening upon yield, followed by large strain hardening. The localized deformation patterns depend on the morphology of the blend and on the triaxiality of the macroscopic stress state. Low triaxiality tends to trigger shear bands across the ligament between neighboring particles, where the orientation depends on the shear direction. At sufficiently high triaxialities, isolated dog-ear shear bands develop which only coalesce at large plastic strains. These dog-ear shear bands in particular lead to regions where craze initiation is likely. If crazing does not intervene, the mesoscopic plastic deformation around voids/cavitated particles lead to significant changes of their size and shape.

References

1. Bucknall, C.B. *Toughened Plastics*, Applied Science Publ., London, 1977; pp 177–179.

2. Donald, A.M.; Kramer, E.J. *J. Mat. Sci.* **1982**, *17*, 1765–1772.

3. Okamoto, Y.; Miyagi, H.; Kakugo, M.; Takahashi, K. *Macromolecules* **1991**, *24*, 5639–5644.

4. Cheng, C.; Hiltner, A.; Baer, E.; Soskey, P.R.; Mylonakis, S.G. *J. Mater. Sci.* **1995**, *30*, 587–595.

5. Borggreve, R.J.M.; Gaymans, R.J.; Eichenwald, H.M. *Polymer* **1989**, *30*, 78–83.

6. Steenbrink, A.C.; Janik, H.; Gaymans, R.J. *J. Mat. Sci.* **1997**, *32*, 5505–5511.

7. Steenbrink, A.C.; Van der Giessen, E. *J. Mech. Phys. Solids* **1999**, *in press*.

8. Pijnenburg, K.G.W.; Van der Giessen, E. *in preparation*.

9. Argon, A.S. *Phil. Mag.* **1973**, *28*, 839–865.

10. Boyce, M.C.; Parks, D.M.; Argon, A.S. *Mech. Mater.* **1988**, *7*, 15–33.

11. Pijnenburg, K. G. W.; Steenbrink, A. C.; Van der Giessen, E. *Polymer* **1999**, *in press*.

12. Steenbrink, A.C.; Van der Giessen, E.; Wu, P.D. *J. Mech. Phys. Solids* **1997**, *45*, 405–437.

13. Argon, A.S.; Hannoosh, J.G. *Phil. Mag.* **1977**, *36*, 1195–1216.

14. Sternstein, S.S.; Myers, F.A. *J. Macromol. Sci. Phys.* **1973**, *B8*, 539–571.

15. Estevez, R.; Tijssens, M.G.A.; Van der Giessen, E. *in preparation*.

Chapter 4

A Multi-Level Finite Element Method for Modeling Rubber-Toughened Amorphous Polymers

Han E. H. Meijer, Leon E. Govaert, and Robert J. M. Smit

DPI, Dutch Polymer Institute, MaTe, Materials Technology,
TUE, Eindhoven University of Technology, P.O. Box 513,
5600 MB Eindhoven, The Netherlands (www.mate.tue.nl)

Polymers are intrinsically tough although they can suffer from extreme localization of the deformation. A pronounced softening after yield, followed by a low hardening, promotes this extreme localization. Post-yield behavior is directly related to the polymer entanglement density, thus to the chain stiffness. A number of – rather academic- experiments, using (in order of increasing entanglement density) Polystyrene (PS), Polymethylmetacrylate (PMMA) and Polycarbonate (PC), clearly demonstrate the importance of the intrinsic post-yield behavior on the macroscopic response. Furthermore polymers differ in their capability of sustaining triaxial stresses, yielding cavitation and –ultimately- crazing. This phenomenon causes notch sensitivity of all polymers, including the so-called tough PC. To overcome this problem, the materials should be made heterogeneous. This paper addresses the use of the Multi-Level Finite Element Method to analyze the heterogeneous deformation of two-phase polymer blends. Two important length scales are considered: the heterogeneous RVE (representative volume element) and that of the continuous scale. Analyses like these not only improve our understanding of the phenomena that occur on the different scales, but also give directions towards improvement of existing materials.

Provided that their molecular weight is sufficiently high to form a real physical network, generally 8 to 10 times the molecular weight between entanglements, all polymers are intrinsically tough. Nevertheless they sometimes appear to be brittle, either during simple slow speed tensile testing (e.g. PS) or, alternatively, under low and high speed, notched, loading conditions (e.g. PC), and worldwide much effort is put in circumventing these obvious problems. Generally heterogeneity is introduced by creating a second dispersed phase. Although for semi-crystalline polymers even hard inclusions proved to be effective, provided that their mutual distance was below a critical value such that the anisotropic crystal orientations could be exploited (*1, 2, 3, 4, 5, 6*), in amorphous polymers up to now only rubbers proved to be successful. Introducing a rubbery phase has a drawback - modulus and yield strength decrease. As a consequence, optimization is sought, by changing the final morphology of the systems investigated and by tuning the properties of the dispersed phase. Recently, modeling capabilities, using relatively advanced constitutive equations and finite element methods, have become mature to the point whereas useful trends can be predicted. The fully 3D constitutive equations used today describe in a proper sense the elastic response, the rate and temperature dependent yield, the intrinsic softening and subsequent hardening behavior of the materials investigated, making use of a principally visco-elastic description. Following the original contributions of Boyce et al. (*7, 8*) nowadays multi-mode versions of the compressible Leonov model (describing a fluid), as proposed by Baaijens (*9*), are used to model polymer solids. The relaxation times, or viscosity's, of the distinct modes of the constitutive model are strongly stress dependent and decrease about 18 decades when the stress increases from zero to –only- 40 MPa (*10, 11, 12*). Yield is thus seen as a stress-induced passage of the glass transition temperature, where the secondary bonds loose connectivity. Softening is added in the model using a more or less empirical approach (*13, 14, 15*). The network contribution (stretching of the surviving primary bonds) is represented by an extra spring, placed in parallel. Different expressions are available to describe the hardening, 3 and 8 chain models by Boyce (*16*), full chain models by Wu and Van der Giessen (*17*), whereas, in many cases, a simple Neo-Hookean approach proves to be sufficient (*15, 18*). The final failure of the stretched network is caused by a stress-induced flow (disentangling) or by chain breakage (*19, 20*).

Amorphous polymers differ in their resistance to triaxial stress states. Surpassing a maximum value (estimated to be 40 MPa for Polystyrene, PS; 75 MPa for Polymethylmetacrylaat, PMMA; and 90 MPa for Polycarbonate, PC (*21, 22, 23*) matrix cavitation is induced and ultimately crazing occurs. Moreover polymers differ in a subtle manner in the relative importance of the intrinsic softening, that can lead to catastrophic localizations. It is the purpose of this paper to illustrate this by a number of dedicated experiments on one hand and advanced calculations of the deformation and failure of homogeneous and heterogeneous polymeric systems on the other hand. The paper is organized as follows: First in an experimental approach homogeneous polymers are examined under different loading conditions. Subsequently, numerical investigations on homogeneous and heterogeneous polymeric systems are performed that basically confirm the outcomes of the experiments on homogeneous systems. Finally, validation is sought by tuning the microstructure of specially designed heterogeneous polymer systems.

Experimental results on homogeneous polymers

A breakthrough in the investigations of the response of different polymers upon loading was achieved by applying compression, rather than the classical tensile tests (*24*). Figure 1 contains the well-known tensile response of brittle PS, intermediate PMMA and tough PC (Figure 1a) as compared to their compression behavior (Figure 1b). The compression tests tend to promote a homogeneous deformation. These tests were stopped when the samples showed barreling. The maximum draw ratio, as determined by the maximum network deformability (*25, 26, 27, 28, 29, 30, 31, 32*) was, consequently, not yet reached. The differences of the three polymers tested, PS, PMMA and PC respectively, are, in compression, only minor and one would not easily predict their tensile behavior based on these experiments that reflect their intrinsic response to loading.

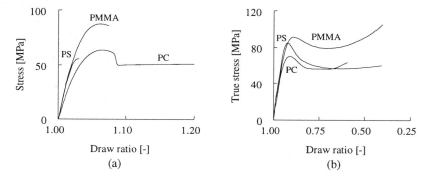

Figure 1. Tensile (a) and compression (b) behavior of PC, PMMA and PS.

Three, rather academic but notwithstanding relevant, tests on homogeneous PS, PMMA and PC have been performed, see (*33*) for further details and a more elaborate discussion. The difference between tensile and compression response was investigated under superposed pressure, under high temperatures and after mechanical rejuvenation. A superposed hydrostatic pressure exceeding 0.4 kbar was shown to be sufficient to suppress crazing (*34*), since the criterion of maximum triaxial (dilative) stress (40 MPa for PS) is never reached. Not surprisingly, the brittle tensile behavior of PS has completely disappeared and a deformation behavior like that achieved in compressive loading is obtained. Similar arguments hold for PMMA, which also completely loses its apparent brittleness and defect sensitivity upon applying a superposed pressure.

The second experiment increases the test temperature, finally approaching the glass transition temperature. Modulus and yield strength, as measured in compression, of course decrease, demonstrated for PS and PMMA in Figures 2a and 2b, respectively. More important is, though, the difference observed between the intrinsic deformation behavior of the two polymers. For PS, Figure 2a, the post yield behavior is only slightly altered and softening is in absolute terms hardly influenced by the test temperature, while for PMMA, Figure 2b, a complete different behavior is observed. Softening gradually disappears upon increasing the temperature.

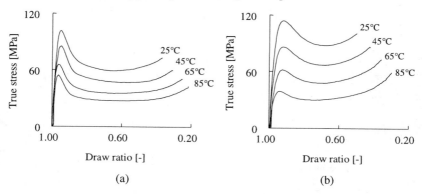

Figure 2. Compression response of PS (a) and PMMA (b) at elevated temperatures and a strain rate of 10^{-2} s^{-1}.

As could have been expected from Figures 1a and 1b, inspecting the compression behavior of PC versus PS and PMMA and the consequent differences in tensile behavior, these differences in the change in intrinsic compression behavior between PS and PMMA upon increasing the testing temperature are of course clearly reflected in their tensile behavior, see Figures 3a and 3b, respectively.

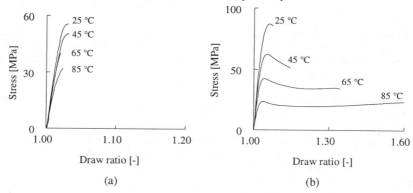

Figure 3. Tensile response of PS (a) and PMMA (b) at elevated temperatures at a strain rate of 10^{-2} s^{-1}.

Whereas PS embrittles further upon increasing the temperature, PMMA becomes ductile. However, upon decreasing the testing speed, also for PS differences are found, see Figures 4a and 4b. Testing at 10^{-3} s^{-1} shows ductile behavior at 85 ^{0}C, while a further decrease of the strain rate to 10^{-4} s^{-1}, yields ductile behavior already at 65 ^{0}C.

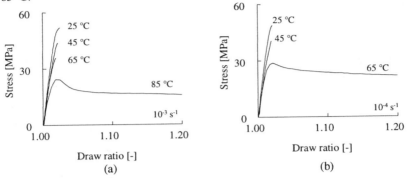

Figure 4. Effect of strain rate on the tensile response of PS at elevated temperature;(a) 10^{-3} s^{-1}, (b) 10^{-4} s^{-1}.

All results of the influence of testing temperature and speed can be summarized in one figure. Figure 5 plots the yield drop observed in a compression test, defined as the difference in the maximum value of the stress-strain curve, the yield point, and the minimum value after yield, caused by intrinsic softening, versus the strain rate applied. Straight lines are found, shifting to lower values at higher temperatures. Above a yield drop of approximately 25 MPa, brittle fracture is observed, below this value ductile behavior is found. This so-called critical value of the yield drop, found at least for the test geometry used, clearly illustrates the importance of the details in post yield behavior of polymers, more specific the overruling influence of the intrinsic softening. This is further demonstrated via a number of intriguing mechanical rejuvenation experiments, our third test.

Thermal pre-conditioning or mechanical pre-deformation have a significant effect on the macroscopic deformation behavior of amorphous polymers. Cross and Haward observed that quenched samples of PVC showed uniform deformation in tensile (*35*), whereas annealed samples showed necking. Bauwens (*36*) prevented necking in PVC by a preceded alternating bending procedure applied on the sample. This this was also shown to be the case for PC, since G'Sell obtained uniform deformation in simple shear, after plastic cycling (*37*, see also: *38* and *39*). Ender and Andrews performed cold drawing experiments on pre-deformed PS (*40*). We redesigned these procedures, in order to be able to measure the intrinsic material behavior. First the results of PC will be discussed. Figure 6a shows the dramatic change in intrinsic behavior of PC after mechanical rejuvenation by twisting

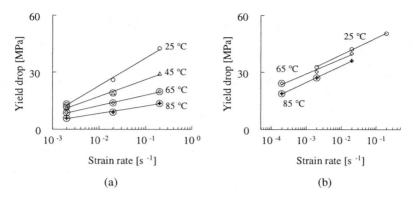

Figure 5. Stress drop after yield as a function of strain rate at different temperatures for PMMA (a) and PS (b). Circled samples are tough.

cylindrical PC samples at room temperature over 720 degrees and back. The softening has been completely removed and tensile and compression results are comparable. After yield, the original hardening curve is recovered, indicating that the network structure has been unaltered by the rejuvenation test and no measurable molecular orientation resulted.

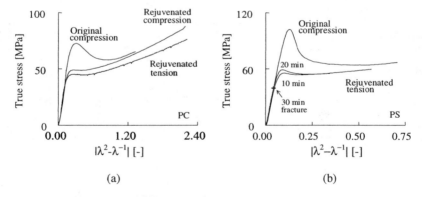

Figure 6. Influence of mechanical rejuvenation on compression and tensile behavior of PC (a) and PS (b).

In contrast to the standard PC samples, during tensile testing of the rejuvenated samples no macroscopic necking was observed and the deformation proceeds homogeneous over the total sample length, until final failure occurred. However, with ongoing time after rejuvenation, the yield stress gradually increases and the subsequent strain softening returns, due to volume relaxation or physical aging. After 10 days, yield stress and strain softening have recovered considerably. Full recovery takes in the order of three weeks. PS and PMMA crazed, during twisting and were, consequently, mechanically rejuvenated in a different way. Of both materials, tensile testing bars were cold rolled on a slow speed hand-controlled two roll mill (diameter of the rolls was 45 *mm*), reducing the thickness to 68%. As a result, the width and length of the tensile bars increased. The rolled PMMA tensile bars show a similar effect while here the recovery time is in the order of 3 days. No results are shown, since we concentrate on the more extreme example of PS. Figure 6b shows the tremendous influence of the mechanical rejuvenation procedure. Macroscopic softening is, comparable to PC and PMMA, completely absent, the material is ductile in tensile, no crazes occur –indirectly confirming that crazing should be preceded by localized yield (*19*)- and the original hardening curve is almost obtained again. The last demonstrates that, although more than in the twisting case, during rolling apparently no significant molecular orientation was induced. The time effect of recovery of strain softening of PS is spectacular. Tests were performed at 10, 20, and 30 minutes after rolling. In the two first tests, some non-critical recovery of the yield stress and strain softening is observed, while at 30 minutes after pre-deformation, for PS brittle fracture is found again. (Especially this influence of pre-deformation on the mechanical behavior in tensile of PS is a nice classroom example. After rolling, knots can be made in the tensile test bar at the beginning of the lecture. Leaving the knotted bar on the desk, during further explanation, it can be broken in a complete brittle way, at the end of the lecture.)

The post-yield behavior of polymers is decisive for their macroscopic response, brittle or ductile. Intrinsic softening causes localization of the deformation. Too strong yield drop leads to catastrophic localization and early failure. In Figure 5 this critical stress drop after yield was observed of approximately 25 MPa, for the test geometries used. This value is not confirmed by the mechanical rejuvenation experiments, where fracture appears to occur at lower values of the yield drop. Obviously more accurate compression and tensile experiments, after rejuvenation followed by physical aging, are required. We do not doubt that these could be performed, but rather point your attention to the following observation. The time effects during volume relaxation and physical aging, that bring back, after a mechanical rejuvenation treatment, the original yield stress and the intrinsic softening and, consequently, brittleness, are spectacular (3 weeks for PC, 3 days for PMMA and 30 minutes for PS), not understood and unexplored, and deserve much more attention.

Numerical results on homogeneous and heterogeneous polymers

Numerical predictions of the deformation and failure of (heterogeneous) polymeric systems was long hampered by the combination of three problems: the difficult intrinsic materials response, the generally complex microstructure and the unsolved problem how to couple the events in the microstructure to the macrolevel and vice versa. Applying the above-mentioned constitutive equations, using material parameters as determined in compression tests, solved the first problem. The second problem has been overcome by applying a detailed finite element calculation on the microscale, i.e. on RVE (representative volume element) level, which should be both: randomly stacked and large enough to be representative (*41*). Finally, introducing a robust, but computationally expensive, MLFEM (multi-level finite element method) analysis (*42, 43*) recently the third problem has been solved. The numerical simulations strongly support the statement that it are the subtle differences in post-yield behavior (softening followed by hardening) that finally determine whether catastrophic localization occurs, and thus brittle behavior. They discriminate between the different polymers used but, moreover, give directions towards possible solutions. We will show (i) the defect versus notch sensitivity of the homogeneous polymers, both yielding craze formation and brittle behavior and thus making heterogeneous systems necessarily, (ii) the RVE-responses of the heterogeneous systems, and (iii) the implications of this behavior for the macroscopic response.

First homogeneous polymers are investigated; using notched tensile testing as the ultimate test. Under the notch, a minor defect is introduced, a local flaw, induced e.g. by the freshly razorblade cut, often applied. The two extremes, PC and PS, are taken as an example. Their intrinsic behavior as tested in compression, see Figure 1b, is used, be it modeled with only one mode, to save computational effort (*44*). In this one mode model, care was taken to accurately describe the post yield behavior, while in the elastic, pre-yield range some inaccuracy was accepted. Upon straining PS, already at an extremely low macroscopic strain of 0.25%, the maximum triaxial stress criterion of 40 MPa for PS is reached just below the defect. The material will cavitate. This process repeats and a craze is formed, perpendicular to the loading direction. The material breaks in a brittle manner. PC, on the contrary, shows less pronounced localization, survives the critical stress below the defect, given its higher resistance against triaxial stresses (90 MPa for PC), and the stress concentration shifts to below the notch. Nevertheless also there the critical 90 MPa criterion is reached, be it at a somewhat higher macroscopic strain as compared to PS, and again a catastrophic craze will form, perpendicular to the loading direction, see Figures 7a and 7b.

We can conclude that both materials show catastrophic localization. The difference between the two materials is that PS clearly is defect sensitive, while PC is only notch sensitive. This is in accordance with practical experience that –in slow speed unnotched tensile testing- PS clearly behaves brittle, the intermediate PMMA reaches the yield point, provided that the samples are carefully polished, while PC is tough, see Figure 1a. To solve this apparently general problem, materials must be made inhomogeneous, e.g. by incorporating a softer dispersed phase, usually a rubber. In order to investigate the influence thereof, heterogeneity is introduced in our FEM model and we investigate the response of an RVE, a representative volume element. In order to be representative, the RVE should be large enough. Simulations have taught us that the total RVE response does not significantly change in different simulations using independently constructed RVE's, when circa 300 randomly stacked holes are introduced, see (*41, 45*). We used holes as a first representation of a dispersed rubbery phase with a modulus of zero MPa. We used a random stack, since this proved to be decisive in the RVE's final response. The RVE's are constructed by cross sectioning a cube filled with spheres with a plane and, subsequently, meshing this plane. Six different volume fractions holes were introduced of 1%, 2.5%, 5%, 10%, 20% and 30% respectively. Upon loading these RVE's in plain strain with periodic boundary conditions (see below), using the constitutive equation for PC, typical shear band formation and cooperative deformation is found, see Figure 8, that compares well with macroscopic tests on a 0.1 mm perforated film of PC (*30, 46*).

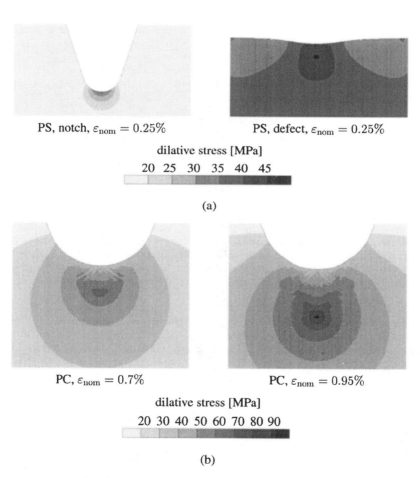

PS, notch, $\varepsilon_{\mathrm{nom}} = 0.25\%$ PS, defect, $\varepsilon_{\mathrm{nom}} = 0.25\%$

dilative stress [MPa]

20 25 30 35 40 45

(a)

PC, $\varepsilon_{\mathrm{nom}} = 0.7\%$ PC, $\varepsilon_{\mathrm{nom}} = 0.95\%$

dilative stress [MPa]

20 30 40 50 60 70 80 90

(b)

Figure 7. Dilative stress under the defect and under the notch for PS (a) and PC (b)

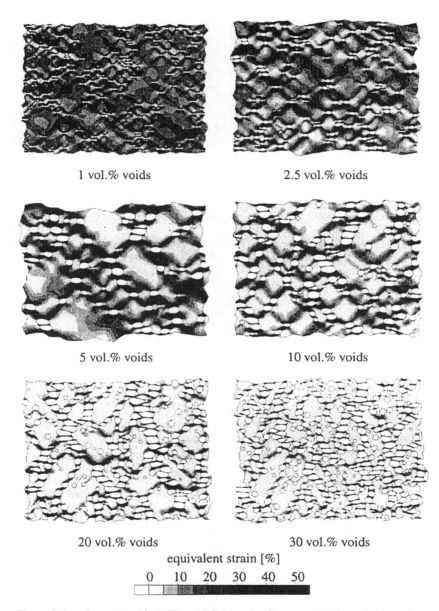

1 vol.% voids   2.5 vol.% voids

5 vol.% voids   10 vol.% voids

20 vol.% voids   30 vol.% voids

equivalent strain [%]

0   10   20   30   40   50

*Figure 8. Local strain inside RVE's of PC upon loading to ca. 20% nominal strain;
volume fraction voids from 1% to 30%*

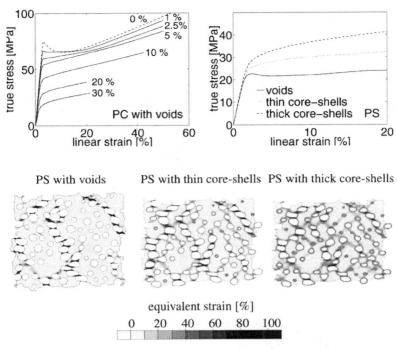

Figure 9. Stress-strain response of the RVE's with different volume fraction voids, for PC (top-left) and for PS (top-right, only the 30 vol.%), and the RVE responses of PS with voids and thin and thick rubbery shells (bottom).

Interesting is the mechanical response of the different RVE's, see Figure 9. Whereas pure PC softens after yield, the addition of voids –apart from the trivial lowering of the modulus and yield stress- makes the softening completely disappear, clearly at least at already 5% volume fraction of voids.

A closer investigation reveals that sequential yielding is the cause of this behavior. Figure 10a presents in a black-white picture these details demonstrated by using the 30-vol.% example. The light grey color indicates those parts that are still only elastically loaded (below yield); black represents all parts that passed the yield point of pure PC. Yield naturally starts at positions of the highest stress (where the ligament between two voids accidentally was small). After hardening, yield is initiated in some other area of the RVE. As a result, yielding gradually spreads out over the total RVE. This is requested for overall tough behavior.

Interesting is now to investigate how PS will perform in an identical RVE. For this purpose, we just change the constitutive behavior from PC to PS, and repeat exactly the same calculation. The results are presented in Figures 9 and 10b.

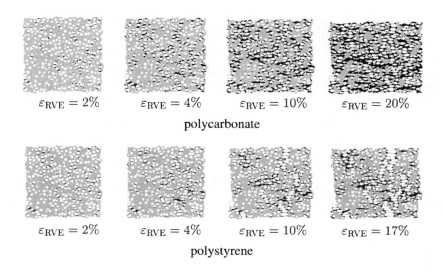

$$\varepsilon_{\text{RVE}} = 2\% \qquad \varepsilon_{\text{RVE}} = 4\% \qquad \varepsilon_{\text{RVE}} = 10\% \qquad \varepsilon_{\text{RVE}} = 20\%$$

polycarbonate

$$\varepsilon_{\text{RVE}} = 2\% \qquad \varepsilon_{\text{RVE}} = 4\% \qquad \varepsilon_{\text{RVE}} = 10\% \qquad \varepsilon_{\text{RVE}} = 17\%$$

polystyrene

Figure 10. Local deformation inside RVE's of PC (a) and PS (b); grey elastic, black plastic.

Compared to PC, in the initial stages of loading, seemingly identical results are obtained. However, the more pronounced softening and less pronounced hardening of PS, causes the ligaments to deform such far, that they hardly can bear any load. They do become so extended that no force can be transmitted towards other regions of the RVE rather than its direct neighbors. Consequently, the bulk of the RVE remains only elastically loaded (and thus 'grey'). A pronounced localization of the deformation is the result and, despite the high breaking stress of the ligaments, they finally will fail. Apparently, PS flows easily after localization, such that it must locally fail, despite of its intrinsic larger drawability. Given its less entangled network structure as compared to PC, yielding for PS involves more molecular orientation in the stretched ligaments and a potentially higher stress to break. It is not the stress bearing capability of the stretched ligaments, but their lack of load bearing capability (stress times cross sectional area) that causes the brittle behavior of PS. Consequently, we must support the filaments during stretching and locally add more stiffness, i.e. resistance to large deformations. This can be realized by filling the voids with e.g. a rubber. We have chosen for a pre-cavitated rubber (yielding a heterogeneous structure that was needed anyway to relieve the triaxial stress state that causes a catastrophic craze, even under a defect) with a thin and a thick shell and a relatively high modulus of 300 MPa. The rubber supports the filaments at both (in our two dimensional plane strain calculations) sides, basically putting two elastic springs in parallel to the filament. As a result, the local hardening of the system is increased. Both measures (thin and thick shells) proved to be sufficient and yield events finally spread out over the whole RVE. This behavior is of course reflected in the RVE's overall mechanical response, that still show softening after yield for the voided PS, in contrast to the voided PC, while in the pre-cavitated rubber filled PS the softening disappeared completely, see Figure 9 and (44).

We will now investigate what the influence is of the RVE responses to the macroscopic response. To couple the two length scales involved, that of the RVE and that of the continuum, macroscopic, scale, is not a trivial issue. Many attempts were tried, including different homogenization concepts. One of the most successful, computationally robust but expensive, methods –proposed in our laboratory by Robert Smit- is the so-called MLFEM, the multilevel finite element method, see (41, 42, 43, 44, ,45). Its principle is simple. Macroscopically, the material is meshed using a finite element method. In the integration points of every element (4 per element) we descend one level down to the RVE, that is considered to be locally periodic, implying that its neighbors deform identically. The macroscopic continuum is loaded by putting an incremental strain on its boundaries. The local strain as calculated in the integration points of every element (using an estimation of the materials constitutive behavior) is transferred to the boundaries of the RVE. Subsequently, the RVE averaged stress is calculated, by performing a finite element calculation on RVE level, as explained in the preceding section. This stress is transferred to the macro-mesh, where stress equilibrium is sought, by solving the momentum equation. Iterations with alternating (imposing local strain, resulting in the averaged local stress, yielding overall macroscopic stress equilibrium) analyses on the macro- and micro-level continue, until convergence is reached. Accordingly, the sample is loaded with the next incremental strain step, and the procedure repeats. Basically, this method uses the RVE analysis as a, non-closed form of the, constitutive equation of the heterogeneous system. Neighboring (each in its own periodic) RVE's, only are connected on the macro-level, via the stress balance, thus assuming that between the RVE's in two neighboring integration points a large number of, in deformation gradually changing, RVE's are present. The results of the analyses of unnotched tensile bars are not reproduced here (see 43 for further details) but they show that –in accordance with experiments- macroscopic homogeneous deformation is found, if the RVE response shows no more softening. The results of the notched tensile bar are summarized in Figures 11a and 11b.

Explaining the findings of Figure 12, we should recall that we already knew that homogeneous PS is defect sensitive, while PC is notch sensitive. In both cases the critical triaxial stress state is reached (either under the defect or under the notch in our present test configuration), and a catastrophic craze, perpendicular to the loading direction results for both materials, be it at different macroscopic strains. Consequently, in the macroscopic response almost no strain to break and a negligible plastic deformation is found. For 30% voided PC, the deformation spreads over a large part of the notched tensile bar. No failure criterion is met, the material behaves tough. This might explain why, already at low volume fractions dispersed rubber (see e.g. Van der Sanden et al. (46), who used 5-vol.% 200 nm sized non-adhering core shell rubbers) the notch sensitivity of PC in practice completely disappears. For 30% voided PS, no interesting improvement is found and the material starts to fail due to a too large deformation in the ligaments of the RVE (local remeshing proved to be necessary and the calculation was stopped). In contrast, the 30% pre-cavitated rubber filled PS shows tough behavior, since the deformation starts to spread in a wide area under the notch and calculations were stopped, without hitting any critical stress state, when remeshing proved to be necessary.

Polymers fail due to their inability to resist triaxial stress states. As a consequence in "brittle" PS, under a defect, but also in "tough" PC, under a notch, the inevitable local yield followed by softening leads to matrix cavitation and catastrophic craze formation. Brittle behavior results. Heterogeneity introduced to –at least partly- overcome this problem proves to be effective. The favorable post-yield behavior of PC (not too pronounced intrinsic softening combined with an effective network hardening modulus that starts, given the limited network drawability, in an early enough stage of the deformation) makes that for this material, already simple measures prove to be sufficient. Adding 5-vol% voids, makes softening to completely disappear at the RVE level (inherently due to sequential yielding and sufficient ligament hardening), causing homogeneous deformation and, consequently, overall tough behavior on the macro-level. For PS, the intrinsic network deformability is that large that catastrophic localization is unavoidable and overall brittle behavior is the result. In order to overcome this problem, the deforming ligaments need to be supported. The voids should be filled with a rubber, be it a pre-cavitated rubber since critical triaxial stress states must always be avoided, even under high speed, notched, testing conditions at low temperatures. The supporting roll of the well-adhering rubber shell yields sufficient hardening and PS can be made tough. It should be emphasized that in all analyses presented so far, no explicit length scales were introduced. This issue clearly needs more attention.

Validation

Both, the experimental study to the influence of the post-yield behavior on the macroscopic tensile behavior of homogeneous amorphous polymers as well as the numerical approaches to address the scale problem and allow for the analysis of heterogeneous polymer systems, basically resulted in some relatively simple but straightforward conclusions. Polymers are intrinsically tough, given their strong covalent bonds that form a strain hardening network structure, combined with their weak secondary bonds that reach the glass transition under tension. They apparently differ in their post-yield behavior. In order to overcome defect-, or notch-, sensitivity, polymers should be made heterogeneous and –in an analysis- the RVE studied should both be irregularly stacked and large enough. Most important, however, is that all different measures that improve the mechanical behavior should -in the end- be explained by either (locally) decreasing the polymers yield stress or – alternatively-

64

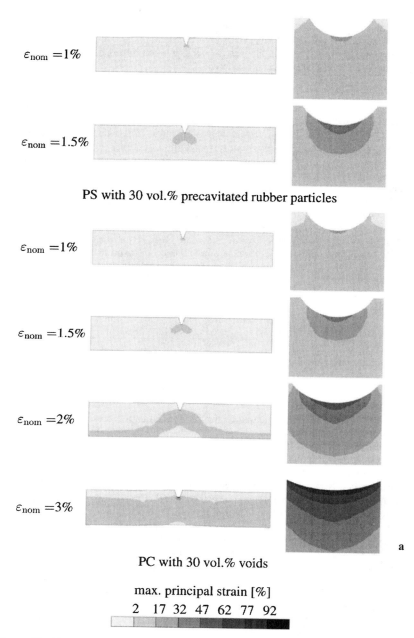

$\varepsilon_{\text{nom}} = 1\%$

$\varepsilon_{\text{nom}} = 1.5\%$

PS with 30 vol.% precavitated rubber particles

$\varepsilon_{\text{nom}} = 1\%$

$\varepsilon_{\text{nom}} = 1.5\%$

$\varepsilon_{\text{nom}} = 2\%$

$\varepsilon_{\text{nom}} = 3\%$

a

PC with 30 vol.% voids

max. principal strain [%]

2   17 32 47 62 77 92

Figure 11. Macroscopic stress-strain behavior (b) and, for the tough materials, the local strains (a) for notched tensile test specimen of PC, PS, voided PC, voided PS and rubbery shell filled PS.

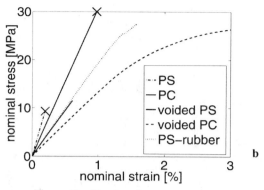

Figure 11. *Continued.*

increasing the polymers hardening modulus. Both measures decrease the yield drop caused by intrinsic softening, and thus prevent the localization to become catastrophic. Well-known measures that have proved to be effective, like cross-linking PS, addition of plasticizers to e.g. PVC or PS, the introduction of heterogeneity and even rubber modification of e.g. PS, are more or less easily explained in these terms.

The existence of an absolute value of a critical ligament thickness between inclusions, below which polymers behave tough, is however not explained yet, since mechanical analyses, including those employing the MLFEM method, are principally scale independent. Consequently, they can not deal with the absolute length scale dependence found in experiments, like the critical ligament thickness for semi-crystalline *(1, 2, 3, 4, 5, 6)* and amorphous *(30, 31, 32, 46)* polymers. In the first, the influence of trans-crystallinity, yielding orientation of the crystallization and, subsequently, anisotropy of the crystal properties (locally decreasing the yield stress) is fully explored. The length scales involved are related the crystal dimensions and the maximum transcrystallinity distance thus yielding a critical distance between either (easy cavitating) soft or (easy debonding) hard fillers that ranges from typically 0.3 μm for Polyamide (PA) to 0.6 μm for Polyethylene (PE). For amorphous polymers similar effects are not found, although their configuration close to walls will not be random. What has been found in non-adhering core-shell rubber modified amorphous polymers is a critical ligament thickness that is network density dependent, and ranges from 0.05 μm for PS to 3 μm for PC *(30, 31, 32, 46)*. Originally, we tried to explain the phenomenon of a critical ligament thickness between non-adhering inclusions in amorphous polymers, via a Griffith type of approach *(32)*. The length scale entered the solution by comparing the (cross sectional) breaking energy of a strained filament with the (volumetric) stored elastic energy in its direct surroundings. Drawback of this approach remains that it is unclear how the polymer knows its breaking energy (and compare that to the stored elastic energy) before it is actually broken. In (*32*), Kramer used a different approach to explain the experimental findings, by introducing an absolute length scale via the decisive influence of the surface tension. Basically he stated that if the absolute value of a ligament is lower than the thickness of three fibrils in a craze (order 30-50 *nm*), no crazing can occur.

In terms of the present analysis, the post-yield behavior of a homogeneous or heterogeneous material is a crucial feature. Stabilization is only possible by a softening reduction or hardening improvement. Consequently, either the yield stress or the strain hardening modulus should be influenced by the absolute size of the materials microstructure. Smit (*44, 47*) tried to explain the experimental findings along this line of thinking. An enhanced mobility of polymer segments near a surface or interface cause the yield stress and strain softening to be reduced. The critical ligament thickness found for PS (order 50 *nm*) is in the same order of the radius of gyration of PS molecules, and a lower yield stress could therefore easily result. Alternatively, if the localized (shear) deformation zone becomes smaller than the entanglement distance of a polymer chain (9.6 *nm* for PS), than only part of the chain is strained, by the stress-induced passage of the glass transition temperature (yield), while the remainder of the chain stays in the glassy state. This effectively results in a quasi-decrease in the molecular weight between entanglements, and thus in a larger hardening modulus. This situation is, again for PS, reached when the inclusions are in the order of 30 *nm*, since the thickness of shearbands formed around inclusions, as calculated by continuum mechanics (still valid on this scale?) are in the order of 15% of the size of the inclusions. Simulations showed that both explanations could –in principal- explain the experimental findings. However, molecular dynamics calculations are needed to get more confidence in these statements.

The validation of the predictions following from the modeling analysis of heterogeneous polymer systems, finally, follows this same line. Using polymerization induced phase separation, meanwhile controlling the coalescence process, special morphologies on the 0.03 μm scale are realized using PS and PMMA as the continuous phase (*48, 49, 50, 51, 52, 53*). Reasons to try to create these special morphologies were inspired by the the experimental findings that the brittle to ultimate-tough transition in PS (defined as an increase in its macroscopic strain to break to at least >150%) is prescribed by the interparticle distance and, consequently, shifted from the addition of >50 vol.% non-adhering core shell rubbers to only >30 vol.%, when the size of the dispersed phase decreased from 200 *nm* (*30, 31*) to 80 *nm* (*48*). Only one example will be given here, demonstrating the synergistic effects found in semi-IPN's of PMMA-aliphatic epoxy structures (also found, but not reproduced here, in real-IPN's and co-polymers of the same system), see Figure 12.

The lower volume fractions of these systems (our original target area) still show poor mechanical properties. This proved to be due to an incomplete phase separation process. However, from 30 vol.% and upwards, the tensile toughness of the systems increased tremendously, see Figure 12, clearly demonstrating its origin in the large deformation capability of the continuous PMMA phase (as compared to the relatively brittle rubber phase). This interesting behavior is, however, not maintained during impact testing. Careful experiments, making use of time-resolved X-Ray scattering

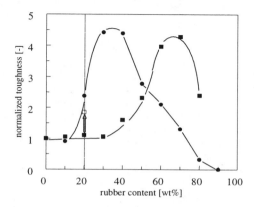

Figure 12. Tensile toughness (unnotched; ●) and impact toughness (notched; ■) of semi-IPN's of PMMA-aliphatic epoxy. The arrow denotes the results of impact testing on pre-cavitated samples of the semi-IPN (□) (50, 52)

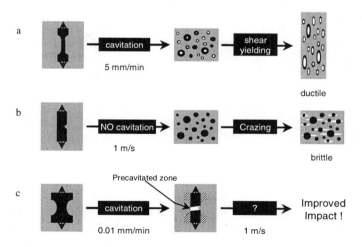

Figure 13. Schematic representation of the relation between mechanical tests, the mode of microscopic deformation, determined via in-situ SAXS measurements, and the macroscopic toughness of 80/20 PMMA/epoxy systems (51, 53).

at the synchroton lines in Grenoble and Daresbury, showed that this was related to the lack of cavitation capability of the dispersed rubbery phase, upon increasing the deformation rate of testing, see Figure 13. Pre-cavitation in low speed tensile loading, followed by impact testing at high speeds, showed the expected improvement, see the arrow in Figure 12.

Discussion

Experimental and modeling efforts on homogeneous and heterogeneous amorphous polymers both emphasize the overruling influence of the materials intrinsic post-yield behavior. In order to obtain ultimate-tough materials, the softening, expressed in the stress drop after yield, should disappear, e.g. on the RVE (representative volume element) level. For different polymers, with their own intrinsic softening (and hardening), consequently with their own yield drop, different measures have to be taken to control the microstructural dimensions and properties, and consequently the RVE's response. For the most challenging material, PS, at present in our laboratories, new routes are being explored to create the "optimal" morphology. They include the use of diblock copolymers, that are selectively miscible in the monomer styrene. Micelles with a size –typically 30 *nm*- that directly depends on the size of the blocks of the diblock, are automatically formed, provided that the size, and inmiscibility, of the other of the blocks is sufficient. The next step is the polymerization of styrene, making PS the continuous phase, meanwhile changing the quality of the solvent and inducing phase separation at the interfaces of the micelles. This route, principally, provides all tools to tune not only the size, but also the properties of the microstructure that consist of a particular type of core-shell rubbers. Its size should be 30 *nm*. It should combine an easy cavitating core, thus a low entanglement density by either use a decreased molecular weight or increased chain stiffness (with modulus typically < 3 MPa), combined with a –for a rubber- relatively high modulus shell (around 300 MPa) a PS matrix of 3000 MPa. An easy to remember target given the magic series of orders of the number 3. Results, hopefully positive, will be reported in the near future.

Acknowledgements: This experimental work was done by Ilse van Casteren, Harold van Melick and Bernd Jansen. The research was supported by the Dutch Technology Foundation (STW), Grant-number EWT.3766 and by the Dutch Polymer Institute (DPI), Grantnumber 163 and 164.

References

1. Muratoglu, O.K., Argon, A.S., Cohen, R.E. and Weinberg. M, *Polymer,* **1995**, *36,* pp. 921-930

2. Muratoglu, O.K., Argon, A.S., Cohen, R.E. and Weinberg. M, *Polymer,* **1995**, *36,* pp. 4771-4786

3. Muratoglu, O.K., Argon, A.S. and Cohen, *Polymer,* **1995**, *36,* pp. 2143-2152

4. Bartczak, Z., Argon, A.S., Cohen, R.E., and Weinberg, M., *Polymer*, **1999**, *40*, pp. 2331-2346
5. Bartczak, Z., Argon, A.S., Cohen, R.E., and Weinberg, M., *Polymer*, **1999**, *40*, pp. 2347-2365
6. Bartczak, Z., Argon, A.S., Cohen, R.E., and Kowalewski, T., *Polymer*, **1999**, *40*, pp. 2367-2380
7. Boyce, M.C., Parks, D.M. and Argon, A.S., *Mechanics of Materials*, **1988**, *7*, pp. 15-33
8. Boyce, M.C., Parks, D.M. and Argon, A.S., *Int. J. Plasticity*, **1989**, *5*, pp. 593-615
9. Baaijens, F.P.T., *Int. J. Numer. Methods Engrg.*, **1993**, *36*, pp. 1115-1143
10. Tervoort, T.A., *PhD thesis TUE*, Eindhoven The Netherlands, **1996**, pp 1-139
11. Tervoort, T.A., Klompen, E.T.J. and Govaert, L.E., *J. Rheol.*, **1996**, *40*, pp. 779-797
12. Tervoort, T.A., Smit, R.J.M., Brekelmans, W.A.M. and Govaert, L.E., *Mech. Of Time Dep. Mat.*, **1998**, *1*, pp. 269-291
13. Hasan, O.A., Boyce, M.C., Li, X.S. and Berko, S, *J. Polym. Sci: Part B: Polymer Phys.*, **1993**, *31*, pp. 185-197
14. Timmermans, P.H.M., *PhD thesis TUE*, Eindhoven, The Netherlands, **1997**, pp. 1-106
15. Govaert, L.E., Timmermans, P.H.M. and Brekelmans, W.A.M., *J. Eng. Mat. Techn.* **1999**, submitted
16. Arruda, E.M. and Boyce, M.C., *Int. J. Plasticity*, **1993**, *9*, p. 697
17. Wu, P.D. and Van der Giessen, E, *Int. J. Mech. Sci.*, **1993**, *35*, pp. 935-951
18. Haward, R.N., *Macromol.*, **1993**, *26*, 5860-5869
19. *Crazing in Polymers;* Kramer, E.J., in *Adv. In Polymers Sci.*, ed. H.H. Kausch, Springer-Verlag, **1983**, pp 1-65
20. *Fundamental processes of craze growth and fracture;* Kramer, E.J., Berger, L.L., in *Adv. In Polymers Sci.*, ed. H.H. Kausch, Springer-Verlag, **1990**, pp 1-68
21. *Crazing and fracture of polymers;* Narisawa, I. And Yee, A.F., in: *Structure and Properties of Polymers*, E.L. Thomas, Vol. Ed., *Volume 12 of Materials Science and Technology. A comprehensive treatment, VCH Wiley*, **1993**, pp. 699-765
22. Ishikawa, M., Sato, Y., and Higuchi, H., *Polymer*, **1996**, *37*, pp. 1177-1181
23. Plummer, C.J.G. and Donald, A.M., *J. Polym. Sci Polym. Phy. ed.*, **1989**, *27*, pp. 325-336
24. Boyce, M.C., Arruda, E.M. and Jayachandran, R., *Pol. Eng. Sci.*, **1994**, *34*, p. 716
25. Smith, P., Lemstra, P.J. and Booy, H.C., *J. Polymer Science, Part B: Phys. Ed.*, **1981**, *19*, p. 877
26. Lemstra, P.J. and Kirschbaum, R., *Polymer*, **1985**, *26*, p. 1372
27. Donald, A.M. and Kramer, E.J., *J. Polym. Sci., Polym. Phys. Edn.*, **1982**, *20*, p. 899
28. Donald, A.M. and Kramer, E.J., *Polymer*, **1982**, *23*, p. 461
29. Henkee, C.S. and Kramer, E.J., *J. Polym. Sci., Polym. Phys. Edn.*, **1984**, *22*, p. 721

30. Van der Sanden, M.C.M., *PhD thesis TUE*, Eindhoven, The Netherlands, **1993,** pp. 1-182
31. Van der Sanden, M.C.M., Meijer, H.E.H. and Lemstra, P.J., *Polymer,* **1993,** *34,* pp. 2148-2154
32. Van der Sanden, M.C.M., Meijer, H.E.H. and Tervoort, T.A., *Polymer,* **1993,** *34,* pp. 2961-2970
33. Van Melick, H.G.H., Govaert, L.E. and Meijer, H.E.H., *Polymer,* submitted
34. Matsushige, K., Radcliffe, S.V., Baer, E., *J. Mater. Sci.,* **1975,** *10,* 833-845
35. Cross, A. and Haward, R.N., *Polymer,* **1978,** *19,* p. 677
36. Bauwens, J.C., *J. Mater. Sci.,* **1978,** *13,* p. 1443
37. *Plastic deformation of glassy polymers: constitutive equations and macromolecular mechanisms,* G'Sell, C., in H.J. Queen et al. Eds, *Strength of metals and alloys,* Pergamon Press, **1986,** p. 1943
38. Aboulfaraj, M., G'Sell, C., Mangelinck, D. and McKenna, G.B., *J. Non-Cryst. Solids,* **1994,** *172-174,* p. 615
39. Hasan, O.A. and Boyce, M.C., *Polymer,* **1993,** *34,* p. 5085
40. Ender, D.H. and Andrews R.D., *J. Appl. Phys.,* **1965,** *34,* pp. 3057-3062
41. Smit, R.J.M., Brekelmans, W.A.M. and Meijer, H.E.H., *J. Mech. Phys. Solids,* **1999,** accepted
42. Smit, R.J.M., Brekelmans, W.A.M. and Meijer, H.E.H., *Comput. Methods Appl. Mech, Engrg,* **1998,** *155,* p. 181
43. Smit, R.J.M., Brekelmans, W.A.M. and Meijer, H.E.H., *J. Mater. Sci,* **1999,** Part III, submitted
44. Smit, R.J.M., Brekelmans, W.A.M. and Meijer, H.E.H., *J. Mater. Sci,* **1999,** Part I, submitted
45. Smit, R.J.M., Brekelmans, W.A.M. and Meijer, H.E.H., *J. Mater. Sci,* **1999,** Part II, submitted
46. Van der Sanden, M.C.M., De Kok, J.J.M. and Meijer, H.E.H., *Polymer,* **1994,** *35,* pp. 2995-3004
47. Smit, R.J.M., *PhD thesis TUE,* **1998,** Eindhoven, The Netherlands, pp.1-103
48. Jansen, B.J.P., *PhD thesis TUE,* **1998,** Eindhoven, The Netherlands, pp.1-104
49 Jansen, B.J.P., Meijer, H.E.H. and Lemstra, P.J., *Polymer,* **1999,** *40,* pp. 2917-2927
50 Jansen, B.J.P., Meijer, H.E.H. and Lemstra, P.J., *Macromolecules,* **1999,** Part I, accepted
51 Jansen, B.J.P., Rastogi, S., Meijer, H.E.H. and Lemstra, P.J., *Macromolecules,* **1999,** Part II, submitted
52 Jansen, B.J.P., Meijer, H.E.H. and Lemstra, P.J., *Macromolecules,* **1999,** Part III, accepted
53 Jansen, B.J.P., Rastogi, S., Meijer, H.E.H. and Lemstra, P.J., *Macromolecules,* **1999,** Part IV, submitted

Chapter 5

Mechanics of Toughening Brittle Polymers

M. D. Thouless[1,2], J. Du[2], and A. F. Yee[2]

Departments of [1]Mechanical Engineering and Applied Mechanics
and [2]Materials Science and Engineering, University of Michigan,
Ann Arbor, MI 48109

The mechanics of toughening brittle polymers by means of a process zone are presented in this chapter. Non-linear processes that act in a process zone around the crack tip may contribute to toughening of a brittle polymer. However, a consideration of the mechanics involved demonstrates that the toughening can only be achieved once material in the process zone is unloaded in the wake of the crack. The toughness then depends on crack length. This is known as R-curve behavior and must be associated with the fracture of any polymer that is toughened by a process zone. The R-curve effect is shown experimentally in a rubber-modified epoxy, and the importance of the region of material in the wake of the crack (rather than ahead of the crack tip) is demonstrated.

Introduction

The crack-driving force

The fundamental physics for the fracture of brittle materials is described by linear-elastic fracture mechanics (L.E.F.M.). This can be summarized as follows. Consider a block of linear-elastic material that contains a crack. When a load is applied to the body, stresses are induced and elastic strain energy is stored throughout the material. In particular, the crack tip introduces a singularity in the stress field and, consequently, there are very high stresses and strain-energy densities in the region near the crack tip. If the crack grows, material that was in the high-stress region immediately ahead of the crack tip passes into the wake of the crack and releases stored strain energy. Meanwhile, material that was in a low-stress region beyond the crack-tip region, moves close to the crack tip, and stores additional strain energy. The overall rate of change in the elastic strain energy is a balance between energy released by elements of material as they pass into the crack wake, and energy stored by elements of material as they advance into the crack-tip region. Furthermore, the compliance of the system is increased by the growth of a crack. If this change in compliance results in displacements at the points of application of any external loads, then work is done by these applied loads. The difference between the total work done on the body by the external loads and the change in the elastic energy locked up in the body is the energy available to propagate the crack. Formally, the crack-driving force (or energy-release rate) is defined as

$$G = \frac{\partial}{\partial A}\left(W - U\right) \qquad (1)$$

where A is the crack area, W is the work done on the cracked body, and U is the elastic energy stored by the body.

The crack-driving force is the energy available to increase the size of a crack by unit area. It is calculated from equations of linear elasticity using the known loads and geometry (including the crack length). Implicit assumptions are that the crack is sharp and that the material really does obey the laws of linear-elasticity everywhere. When dealing with polymers, of course, the crack will generally not be sharp at a molecular scale, and there may be local regions of material around the crack that deform by non-linear processes such as crazing, cavitation and yielding. However, provided the scale of these events is small enough that they provide a negligible influence on macroscopic observations such as a load-displacement curve obtained in the absence of any crack growth, the concept of a crack-driving force as defined in Equation (1) is an appropriate tool with which to analyze fracture. (If crack growth does occur while obtaining a load-displacement curve, compliance changes caused by the crack growth may introduce substantial non-linearities in the curve. If this is the only cause of non-linearity, the assumptions of L.E.F.M. remain completely valid.)

Even if they have a negligible effect on macroscopic load-displacement curves, local regions of non-linearity around a crack tip do have a very important effect on fracture. The actual energy that is available to propagate a crack may be different from the energy that is calculated from Equation (1), even if the material behaves in a linear-elastic fashion overall. Some of the energy calculated from Equation (1) may be dissipated in the non-linear region, so that the crack tip is "shielded" from being able to use all the energy supplied by the system. This lies at the heart of the concept of toughening brittle materials, and it is convenient to define two different quantities for the crack-driving force. One is the actual energy that can be used to break the bonds at the crack tip; this is designated G_{tip}. It can be considered to be the energy-release rate that the crack "sees". The other quantity is the nominal energy-release rate, or the nominal crack-driving force, G^∞. This is the quantity calculated mathematically based on the assumption that the crack is sharp, that the loads and geometry are known exactly, and that the constitutive properties of the material are uniform throughout the body and are described by the macroscopic load-displacement data.

Toughness

At the crack tip itself, the polymer is held together by molecular chains and atomic forces that provide bonding across the crack plane. As the crack surfaces are separated, this bonding exhibits a characteristic relationship between the cohesive stress and the displacement of the crack surfaces, as illustrated in Figure 1. The peak stress that the cohesive processes can support is designated by $\hat{\sigma}$. This peak stress, and the maximum displacement by which the crack surfaces can be separated before there are no cohesive stresses acting between them, are determined by the intrinsic nature of the polymer and its molecular structure. In a crystalline material, the decohesion process reflects the theoretical cleavage strength of an atomic bond. In polymeric materials, chain scission, disentanglement and pull-out are all expected to contribute to the characteristic stress-displacement curve of the decohesion process. Of particular importance is the fact that the area under the stress-displacement curve represents the intrinsic toughness of the material, which is designated Γ_o:

$$\Gamma_o = \int_0^{\delta_c} \sigma d\delta \qquad (2)$$

where σ is the cohesive stress, δ is the displacement of the crack surfaces during the decohesion process, and δ_c is the critical displacement for complete decohesion. Γ_o is the energy per unit area that must be supplied to the crack tip in order for the decohesion process to proceed to completion, and to permit the crack to grow by unit area. Hence, Γ_o is an intrinsic material property (that will, for polymers, depend on both the temperature and rate at which decohesion is proceeding). To determine whether it is thermodynamically possible for fracture to occur, Γ_o must be compared to the crack-driving force supplied to the crack tip. The condition for fracture is that $G_{tip} \geq \Gamma_o$.

The cohesive stress acts on the region of material surrounding the crack tip. If these stresses are high enough, then non-linear processes such as yield, cavitation or crazing may be triggered in this region. However, since there is a highly triaxial stress state at the crack tip, these processes can be triggered only if the peak cohesive stress, $\hat{\sigma}$, is greater than about 2 to 3 times the critical tensile stress required to initiate the same processes under uniaxial conditions, σ_{ct}. If no non-linear deformation occurs around the crack, then $G_{tip} = G^\infty$, and the condition for crack advance is simply that the nominal crack-driving force must exceed the intrinsic toughness of the polymeric matrix (at the appropriate crack velocity); i.e., $G^\infty \geq \Gamma_o$.

Non-linear deformation in a process zone around the crack has two effects. By considering the stress-strain history of an element of material immediately ahead of the crack tip, one can see that if the element is loaded in a non-linear fashion, the total energy absorbed is less than would have been absorbed had the element remained elastic. (Depending on the details of the mechanics, it is conceivable that this would result in more energy reaching the crack tip, and hence contributing to an apparent embrittlement of the polymer!) However, if energy is dissipated by this non-linear deformation (rather than being simply stored reversibly), a much more important effect comes into play once the crack begins to grow and the strained element passes into the wake. If the element remains elastic, all the absorbed energy is released upon unloading and is available to help the decohesion process. If dissipation occurs, this energy is _not_ available to the crack tip and the actual crack-driving force, G_{tip}, is less than the nominal crack-driving force, G^∞, calculated for the same geometry and applied loads (but based on the assumption of uniform linear elasticity throughout the polymer). The relationship between G^∞ and G_{tip} can then be expressed as

$$G_{tip} = G^\infty - \Delta G \qquad (3)$$

where ΔG represents the reduction in energy-release rate associated with the fact that less energy is released to the crack tip than would be predicted from the global analysis. It is important to appreciate that this difference is only realized once material in the process zone is unloaded.

Fracture occurs when the energy-release rate supplied to the crack tip equals the energy required for complete decohesion, i.e., $G_{tip} = \Gamma_o$. However, the quantity that can be determined experimentally is the value of G^∞ at which the crack is observed to grow. This is the quantity defined as "toughness" or "fracture resistance". A fracture experiment is performed by determining the load required to fracture a specimen of known geometry. The value of the nominal crack-driving force, G^∞, at this critical load is calculated from the mechanics, and is equated to the toughness of the polymer,

Γ. In other words, the measured value of toughness that is quoted for a particular polymer as actually defined by

$$\Gamma = \mathcal{G}^{\infty} \quad \left|\begin{array}{l}\text{calculated from}\\ \text{the conditions}\\ \text{at fracture}\end{array}\right. \tag{4}$$

A comparison with Equation (2) shows that the measured toughness of a brittle polymer is related to the intrinsic toughness by:

$$\Gamma = \Gamma_{o} + \Delta\Gamma \tag{5}$$

where $\Delta\Gamma$ is the energy not released by <u>unloading</u> of material in the crack-tip region. The great strength of L.E.F.M. is that the same argument applies in reverse when it is used to predict failure loads for any other geometry. One does not need to compute the actual energy-release rate at the crack tip, rather one can assume that failure will occur when $\mathcal{G}^{\infty} = \Gamma$ for any geometry (provided the process zone is embedded completely in a large elastic region).

This concept of reduced energy being released to the crack tip from material in a non-linear process zone upon <u>unloading</u> lies at the heart of the physics of toughening brittle polymers. In particular, it should be emphasized that energy "absorbed" in the non-linear process is not a measure of toughening. When material ahead of a crack is loaded, the energy that is supplied to it is either stored (during an elastic process) or dissipated (during a dissipative process). In either case, this energy is not available to cause decohesion at the crack tip. It is only when the crack begins to move, and the material is unloaded, that it becomes important whether energy has actually been dissipated, or whether the non-linear process was actually reversible. Until one tries to unload material one cannot determine whether energy has been dissipated or merely stored. (A flippant, but perhaps useful, analogy to this concept is to consider that paper gains and losses in a stock market can only be realized when the shares are sold!) The frontal process zone needs to pass into the wake of a crack before appreciable toughening can occur. Indeed, a frontal process zone can cause negative toughening - energy dissipated in a process zone ahead of a stationary crack may cause a <u>decrease</u> in the toughness. The shape of the process zone is critical in determining whether there is any toughening or not; full and rather complex mechanics analyses are required to determine whether a stationary crack is slightly toughened or slightly embrittled by a non-linear process zone that develops ahead of it [1-3].

R-curve Behavior

As a crack propagates, material in the process zone unloads and contributes to decohesion by releasing stored elastic energy to the crack tip. Any dissipative process that occurs in the process zone reduces the amount of energy available to the crack tip. The difference between the energy that would have been released had all the material around the crack tip behaved in a linear-elastic fashion and the actual energy that is available is manifested as an increase in toughness. As the crack progresses through the process zone, the difference between these two quantities increases as more material is completely unloaded. This leads to an increase in toughness with crack length, as illustrated in Figure 2. When there is little or no crack advance, and relatively little unloading of the process zone, the toughness of the polymer is approximately equal to the intrinsic toughness of the polymer. (As mentioned above, whether this initial toughness is greater or smaller than the intrinsic toughness of the polymer depends on the precise shape of the non-linear process zone.) As the crack grows, the toughness increases until a steady-state is eventually reached in which

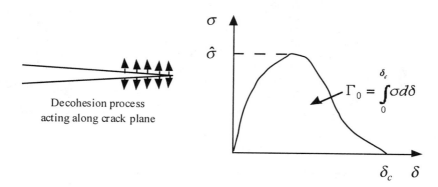

Figure 1. Schematic diagrams of the decohesion process and the characteristic relationship between the cohesive stress and the displacement of the crack surfaces. (Reproduced with permission from reference 18. Copyright 1999.)

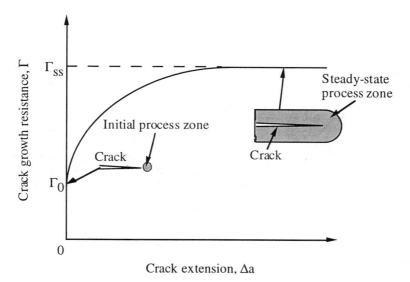

Figure 2. Schematic diagram of an R-curve for materials toughened by a process zone. Γ_0 is the initial toughness, and Γ_{ss} is the steady-state toughness. (Reproduced with permission from reference 18. Copyright 1999.)

material that had originally been in the process zone ahead of the crack tip is completely unloaded in the wake of the crack. This phenomenon is known as "R-curve" behavior, and is a necessary feature of toughening by a process zone. If there is no R-curve behavior then a process zone can not be responsible for toughening - any toughening that is observed must arise from some other process such as a change in the intrinsic toughness of the polymer, or unconstrained blunting of the crack tip in which non-linear deformation extends all the way to the specimen boundaries before any crack growth occurs. (Some confusion is caused by the additional use of the term "R-curve" to denote this phenomenon of blunting and tearing in ductile polymers. It must be emphasized that the phenomena described in this paper apply to polymers that are not ductile, and for which ductility is limited to a small region around the crack that is embedded within an elastic medium.)

While the mechanics required to describe the toughness of a polymer with an evolving process zone is complex, the steady-state can be characterized in a fairly straight-forward fashion. The steady-state toughening is given by

$$\Delta\Gamma_{ss} = 2\int_0^h dy \oint \sigma(y)d\varepsilon \tag{6}$$

where h is the steady-state thickness of the process zone in the wake of the crack, y is the distance from the crack plane, and $\oint \sigma(y)d\varepsilon$ is the energy dissipated by a unit volume of element of material situated at a distance y from the crack plane as it is loaded up to the maximum stress it experiences as it passes through the process zone and is then completely unloaded in the wake (Figure 3). Once the toughening process has reached a steady-state in which an increment of crack advance results in material effectively undergoing an entire load/unload cycle, all the energy absorbed by the process zone is realized as an increase in toughness. Consequently, the expression given in Equation (6) is identical to the results of models [4, 5] couched in terms of energy absorbed in a process zone ahead of a crack. The essential piece of physics missing in these types of models is that the result is for a steady-state toughness only. The proposed mechanisms and physical understanding of the energy absorption are appropriate. However, the crack must grow, and material in the process zone must be unloaded, before this toughness is realized. Predictions of these process-zone toughening models must therefore be compared to measurements of steady-state toughness. It is essential that the full R-curve is obtained during experimental investigations of toughness. During a fracture experiment it is possible for a crack to go unstable during the rising R-curve portion of crack growth. If such an instability occurs, the measured toughness will correspond to the toughness for the crack length at which the instability occurred - this may be less than the steady-state toughness.

Summary

There are basically two ways by which the toughness of a brittle polymer can be increased. The first is to try and increase the "intrinsic" toughness. For example, if the origin of the cohesive stresses that act at the crack tip are understood, it might be possible to tailor the molecular architecture to increase the area under the stress-separation curve, by, for example, increasing the maximum separation at fracture of the crack surfaces. An increase in toughness might result if it were possible to change the decohesion mechanism. This could, perhaps, be achieved by activating a mechanism in which crack advance was associated with cavity growth and coalescence. Alternatively, additional cohesive stresses could be introduced by means of bridging mechanisms such as would be caused by the addition of large particles of rubber [6]. (This description is an acceptable alternative to bridging in the crack wake - it is just a matter of perspective and emphasis.)

Process-zone toughening is introduced by increasing the ratio of $\hat{\sigma}/\sigma_{ct}$ to a level sufficient to trigger non-linear deformation in a local region around the crack tip. This could be done by increasing $\hat{\sigma}$, but a more likely way to achieve this would be to lower the critical stress required to trigger non-linear deformation such as cavitation, crazing or yielding. In general, the toughness of a brittle polymer toughened by a process zone is given by:

$$\frac{\Gamma}{\Gamma_o} = f\left(\frac{\sigma_{ct}}{E}, \frac{\hat{\sigma}}{\sigma_{ct}}, a, \text{constitutive properties of matrix}\right) \qquad (7)$$

Numerical calculations by Tvergaard and Hutchinson [7] and by Wei and Hutchinson [8] indicate that the steady-state toughness is particularly sensitive on the ratio of $\hat{\sigma}/\sigma_{ct}$. Below about 3, the toughness is not substantially greater than the intrinsic toughness. However, once the ratio is above 3.5, the steady-state toughness can be elevated by a factor of more than 20. As this ratio is increased, the size of the process zone increases dramatically, and a previously brittle material can act as a very tough material indeed. Therefore, small changes in the critical stress required to trigger non-linear deformation can have dramatic effects on whether a polymer behaves in a brittle or tough fashion. Controlling this ratio is crucial in any attempt to tailor the microstructure of polymers to achieve toughness. This lies at the heart of the approach used to toughen brittle epoxies by the addition of small rubber particles. As demonstrated by a number of authors, cavitation of the rubber particles in the process zone allows the matrix to deform by yielding more readily than without the rubber [9-12].

Experimental

Most existing models of toughening brittle polymers are process-zone models [4, 5, 13, 14]. However, there has been very little experimental effort in the polymers literature to measure the R-curves that must, by necessity, occur for any material that is toughened by a process zone. A well-characterized model system of a DER® epoxy cured with piperidine and toughened with a CTBN elastomer was used to demonstrate this R-curve behavior [15]. It is well-established that these materials are toughened by a process in which the rubber particles around the crack tip cavitate allowing the matrix to deform by shear yielding [9-12].

The details of the processing are described elsewhere [15]. Double-cantilever-beam specimens were used to measure the toughness. A side groove was machined in the samples to ensure stability of the crack in the central plane of the specimen; experiments without the side groove showed that it had no effect on the results beyond increasing the distance over which the crack propagated along the center-line of the specimen. The tests were run in a quasi-static mode in which the loading was halted immediately after an increment of crack growth was observed. This had the effect of leaving a striation on the fracture surface at the point where the crack arrested. After complete failure, these striations were used to determine crack lengths which were correlated with the appropriate load to determine the nominal crack-driving force. Subsequent experiments in which the crack was allowed to grow continuously and was monitored optically, showed that the quasi-static results were consistent with results obtained at very low crack velocities.

R-curves

Plots of the fracture resistance against crack extension are shown in Figure 4 for the rubber-modified epoxy that contains 10 parts of rubber per hundred parts of resin by weight. It will be observed that the fracture resistance increases from an initial value of about 2 kJ/m^2 to a steady-state value of about 7 kJ/m^2. Two things are of particular note. The first is the unambiguous nature of the R-curve. The second, rather surprising result is that the initial toughness of these composites appears to be considerably greater than the 200 J/m^2 or so that is measured for the pure epoxy. This indicates that the addition of rubber to the matrix may play an additional role beyond simply providing a process-zone mechanism of toughening. It is possible that the second phase may be contributing to an increase in the "intrinsic" toughness; although, what this mechanism might be is not immediately obvious.

The evolution of the process zone can be seen by taking thin sections from the center of the specimens and examining them in transmitted light. Scattering of light from the voids allows the shape of the process zone to be seen clearly, as shown in Figure 5. These process zones appear to be classic examples of the shapes predicted in the mechanics literature. Furthermore, it is of interest to note that the rounded shape is similar to that predicted for dilatation-dominated deformations [16, 17].

The mechanics of R-curves indicates that it is the portion of the process zone in the wake of the crack that provides the dominant contribution to toughness. Therefore, an obvious question is whether the toughness changes when the process zone is removed from the crack wake. This was investigated by a series of experiments in which the process zone was machined away up to pre-determined distances from the surface trace of the crack tip on the side of the samples after a steady-state R-curve was established (Figure 6). The samples were then reloaded and the R-curve re-established. The results for the 10-phr rubber-modified epoxy are shown in Figure 7. (Since the crack had a thumb-nail profile caused by the effects of local plane stress at the free surfaces of the specimen, the amount of material left in the wake is slightly larger than that indicated on this figure. For example, when about zero millimeters of process zone were left at the edge of the sample, there was about 1 mm of process zone behind the crack tip at the center of the sample. This is why the toughness does not drop down to the intrinsic value of the toughness.) Of particular note in this figure is the fact that removal of the wake results in a dramatic drop in the toughness. Upon further crack growth, the R-curve re-establishes itself, and the toughness builds up to the steady state value. This confirms the notion that it is the wake that contributes to the toughness of a process-zone toughened material.

Load-displacement Curves in a Fracture Test

An important consideration for the mechanics of the type discussed in this chapter is that the process zone remains small compared to the dimensions of the sample. Under these conditions, the overall deformation of the material is dominated by the linear-elastic properties of the material outside the process zone. In particular, this means that a load-displacement curve obtained for a body with a non-propagating crack should behave in a linear fashion. If the process zone ahead of a crack grows to too great an extent (*i.e.*, to a sizable fraction of the specimen size) during loading, non-linear processes will start to influence the overall deformation of the material. The load-displacement curve will then exhibit a marked deviation from linearity as

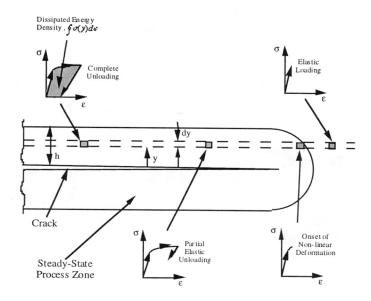

Figure 3. Stress-strain history of an element of material as it passes through a steady-state process zone.
(Adapted with permission from reference 13. Copyright 1986.)

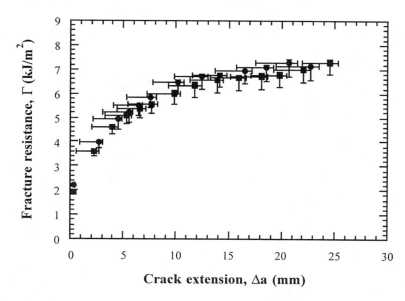

Figure 4. Experimental R-curves for the 10-phr rubber-modified epoxy. The results for two separate specimens are designated by ■ and ● (from [15]).
(Reproduced with permission from reference 18. Copyright 1999.)

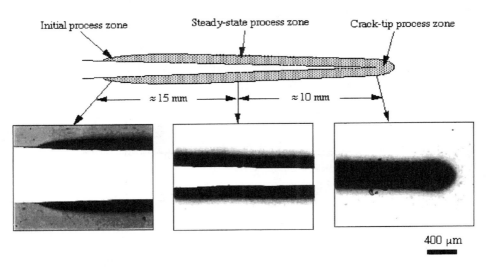

Figure 5. Transmission optical micrographs of the process zone for the 10-phr rubber-modified epoxy (from [15]).
(Reproduced with permission from reference 18. Copyright 1999.)

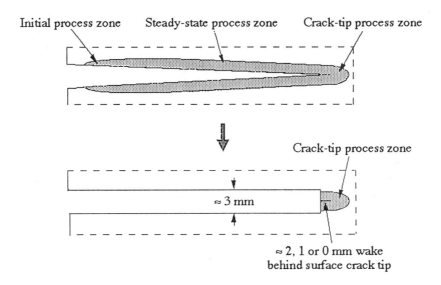

Figure 6. Schematic diagram of the removal of the crack wake.
(Reproduced with permission from reference 18. Copyright 1999.)

the crack blunts and flow occurs in the process zone ahead of the crack. A schematic illustration of the load-displacement curve for such a situation is shown in Figure 8. ASTM standard E399 provides a protocol for determining the critical load for crack propagation. Essentially, this consists of taking a line that has a slope which is 95% of the initial slope of the load-displacement curve, and equating the intersection of the actual displacement curve with this line as the point at which crack propagation occurs. The corresponding load is then used to compute a nominal energy-release rate that is equated to the toughness of the material. The toughness is assumed to be a single-valued quantity in this approach, and no allowance for an R-curve of the type described in this chapter is made.

It can be shown that a load-displacement curve very similar to that illustrated in Figure 8 will be obtained <u>during crack growth</u> for a material that is dominated by linear-elasticity, but which exhibits R-curve behavior. However, it is important to appreciate that, despite superficial resemblances, the non-linearity comes from a totally different source. A load *versus* crack-opening-displacement (measured by a clip gauge at the mouth of a growing crack) curve for a 10-phr rubber-modified epoxy is shown in Figure 9. It would be totally wrong to interpret this curve in the same fashion as Figure 8. Now, the point at which the curve deviates from linearity is the point at which crack growth begins to occur. This can be confirmed by optical observations. Crack growth causes a compliance change (*i.e.*, for a given load, the resultant displacement is larger). The effect of crack length on compliance can be seen from the slope of the unloading curve. The lower slope corresponds to an increased compliance caused by the longer crack length. The rising load indicates that the toughness of the material is increasing. Eventually, when steady-state is reached, the effect of the increasing compliance dominates, and the load drops after reaching a maximum value.

It will be observed from Figure 9 that the unloading curve does not go completely to zero as might be expected at first. There is a residual opening of about 750 μm out of a total displacement of 5 mm. This is caused by residual deformations in the wake of the crack. Cavitation and void growth occur in the process zone. These cause expansion against the elastic matrix, which in turn forces the crack to close. (Although an energy argument is used to describe the toughening process in these materials, it is also possible to describe the toughening process in terms of the closing force on the crack tip.) The effect of this residual compression can be observed by monitoring the crack opening while the wake is machined out as in Figure 6. The results of such an experiment are shown in Figure 10. It will be observed that when the wake is completely removed up to the trace of the crack on the surface of the sample, only 100 μm of residual displacement is left. This is an upper-bound on the deviation from linearity corresponding to the non-linear process zone <u>ahead</u> of the crack, and is a proper measure of the extent to which the polymer deviates from an ideal linear-elastic material. In this case this residual displacement of 100 μm is less than 2% of the total deformation of about 5 mm. This represents the behavior of a material that can be considered to be very linear-elastic.

Conclusions

The most important point made in this chapter is that there are, in principle, two approaches by which a brittle polymer may be toughened. The first would be to manipulate the cohesive processes that act across a crack tip, so as to increase the intrinsic toughness of the polymer. The second approach is to manipulate the

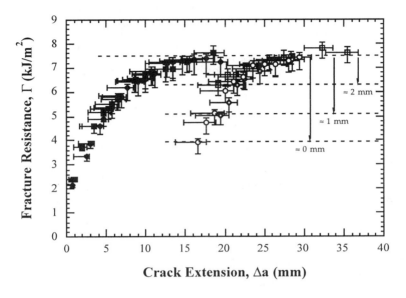

Figure 7. Experimental R-curves for the 10-phr rubber-modified epoxy. The solid symbols(■, ● and ◆) represent the data points obtained before the crack wake was removed. The open symbols (□, ○ and ◇) represent the corresponding data points obtained after the crack wake was removed. The effect of removing the wake up to 2, 1, and 0 mm from the surface trace of the crack is shown (from [15]).
(Reproduced with permission from reference 18. Copyright 1999.)

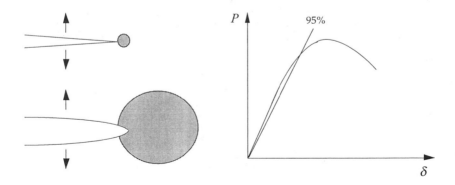

Figure 8. Schematic diagrams of crack blunting associated with extensive plastic flow, and the resultant load-displacement curve which exhibits a significant non-linearity.

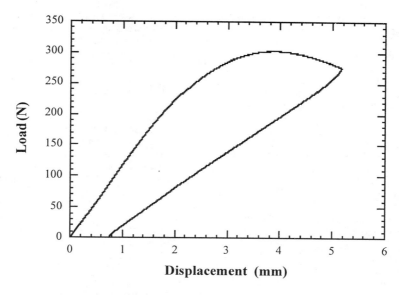

Figure 9. An experimental load-displacement curve for the 10-phr rubber-modified epoxy obtained in a double-cantilever-beam test.
(Reproduced with permission from reference 18. Copyright 1999.)

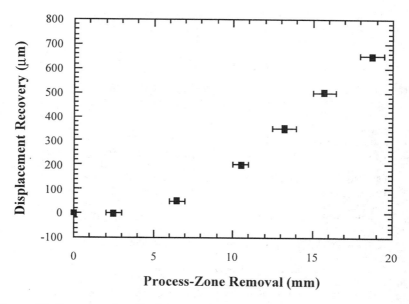

Figure 10. Recovery of residual-opening-displacement at the crack mouth plotted against length of process-zone removed for the 10-phr rubber-modified epoxy.
(Reproduced with permission from reference 18. Copyright 1999.)

microstructure of the polymer so as to make it easier to trigger non-linear dissipative processes around the crack tip. This can be done by increasing $\hat{\sigma}/\sigma_{ct}$ (the ratio of the peak cohesive stress to the critical stress required to trigger non-linear deformation) above about 3. This process zone shields the crack tip from the applied energy-release rate. This is the implicit strategy behind the very successful development of rubber-modified epoxies. The models described in the literature as part of this research capture the essence of this strategy. However, it is important to appreciate that a dissipative process occurring ahead of a stationary crack is not sufficient to ensure toughening. Toughening by a process zone is only realized when the zone extends into the wake of the crack. This results in a toughness that increases from the intrinsic toughness to a steady-state value as the crack grows. This is known as "R-curve" behavior. Such behavior must occur in any polymer that is toughened by a process zone. Furthermore, it is important to obtain complete R-curves when characterizing toughness (and strengths) of such materials; single values of toughness should not be quoted unless it is unambiguous that these are steady-state values.

The extent of toughening is extremely sensitive to the ratio of $\hat{\sigma}/\sigma_{ct}$. Small changes around the value of 3, can have significant effects on the toughness, changing the material from ductile to brittle. Both the intrinsic cohesive processes and the non-linear processes in the process zone may be very rate-sensitive in polymers. If the ratio of $\hat{\sigma}/\sigma_{ct}$ is close to the critical value of about three, small changes in crack velocity that change either $\hat{\sigma}$ or σ_{ct} may cause transitions between ductile and brittle behavior. These rate effects, and accompanying instabilities are currently being studied within the context of the mechanics presented in this chapter.

Acknowledgments

This work was supported by the NSF under grant # CMS-9523078.

References

1. McMeeking, R. M.; Evans, A. G. *J. Am. Ceram. Soc.* **1982**, *65*, 242-246.
2. Budiansky, B.; Hutchinson, J. W.; Lambrapolous, J. C. *Int. J. Solids Struct.* **1983**, *19*, 337-355.
3. Evans, A. G.; Faber, K. T. *J. Am. Ceram. Soc.* **1984**, *67*, 255-260.
4. Huang, Y.; Kinloch, A. J. *J. Mater. Sci.* **1992**, *27*, 2763-2769.
5. Kinloch, A. J.; Guild, F. J. In *Toughened Plastics II: Novel Approaches in Science and Engineering*; Riew, C. K.; Kinloch, A. J. Eds.; Advances in Chemistry Series 252; American Chemical Society, USA, 1996; pp 1-25.
6. Kunz-Douglass, S.; Beaumont, P. W. R.; Ashby, M. F. *J. Mater. Sci.* **1980**, *15*, 1109-1123.
7. Tvergaard, V.; Hutchinson, J. W. *J. Mech. Phys. Solids.* **1992**, *40*, 1377-1397.
8. Wei, Y.; Hutchinson, J. W. private communication
9. Kinloch, A. J.; Shaw, S. J.; Tod, D. A.; Hunston, D. L. *Polymer* **1983**, *24*, 1341-1354.
10. Kinloch, A. J.; Shaw, S. J.; Hunston, D. L. *Polymer* **1983**, *24*, 1355-1363.
11. Yee, A. F.; Pearson, R. A. *J. Mater. Sci.* **1986**, *21*, 2462-2474.
12. Pearson, R. A.; Yee, A. F. *J. Mater. Sci.* **1986**, *21*, 2475-2488.

13. Evans, A. G.; Ahmad, Z. B.; Gilbert, D. G.; Beaumont, P. W. R. *Acta Metall.* **1986**, *34*, 79-87.
14. Argon, A. S. In *Advances in Fracture Research*; Samala, K.; Ravi-Chander, K.; Taplin, D. M. R.; Rama Rao, P. Eds.; ICF7; Pergamon Press: New York, USA, 1989; Vol. 4, pp 2661-2681.
15. Du, J.; Thouless, M. D.; Yee, *A. F. Int. J. Fract.* in press.
16. Rose, L. R. F.; Swain, M. V. *Acta Metall.* **1988**, *36*, 955-962.
17. Yu, C.-S.; Shetty, D. K. *J. Am. Ceram. Soc.* **1989**, *72*, 921-928.
18. Du, J. Doctoral Dissertation. University of Michigan, Ann Arbor, MI, 1999.

Chapter 6

Three-Dimensional Elasto-Plastic Finite Element Modeling of Rigid–Rigid Polymer Toughening Processes

Yiu-Wing Mai and Xiao-hong Chen

Center for Advanced Materials Technology (CAMT), Department of Mechanical and Mechatronic Engineering J07, University of Sydney, Sydney, New South Wales 2006, Australia

Three-dimensional elasto-plastic finite element modelling is developed to simulate toughening processes in rigid-rigid polymer blends. The face-centred cuboidal cell model is provided to account for full inter-particle interactions. The effective constitutive relation is obtained by the method of homogenisation. Local stress concentrations responsible for cavitation, crazing and shear yielding are calculated at different stress triaxiality. The anelasticity and plasticity of second-phase particles have led to enhancement in stress concentrations and reduction in effective yield stress. Cavitation of second-phase particles is shown to play an important role in the rigid-rigid polymer toughening processes especially at high stress triaxiality.

It is well known that rubber modification is a most effective way to improve the fracture toughness of many brittle polymers; but unfortunately it also leads to simultaneous reductions in modulus and strength (*1-3*). Chen & Mai (*4,5,6*) provided a face-centred cuboidal cell model to study the toughening mechanisms in rubber-modified epoxies by 3-D elasto-plastic finite element analysis (FEA) because it is easy to simulate both rubber cavitation and matrix shear banding processes using such a staggered periodic layout. In addition, they combined the face-centred cuboidal cell model with a particle-crack tip interaction model to study the deformation and fracture behavious of rubber-toughened polymers (*7,8*).

A new generation of polymer blends with beneficial combination of toughness, strength and modulus is required to provide a wider range of engineering applications for polymer materials. ABS/PC, PA/PPO, PBT/PC and PEI/PC are typical examples of such blends. Recent introduction of the rigid-rigid polymer toughening concept was originated from the idea of rubber toughening (*9*). It is realised that a rigid polymer matrix can be toughened by the presence of a second-phase rigid polymer as long as the latter is able to act as stress concentrators and help relieve the triaxial constraint. The substantial increase in the fracture toughness of PBT/PC blends comes mainly from the debonding-cavitation process at the PBT/PC interface, which in turn promotes extensive shear deformation in both the PBT and PC; as well as from the craze-stabilising and crack-bridging mechanisms by the PC domains as observed by Wu, Mai & Yee (10).

Unlike ordinary rubber-toughened polymers, both phases of rigid-rigid polymer blends can undergo anelastic and plastic deformation. Sue, Pearson & Yee (9) have modelled the initiation of localised yielding in rigid-rigid polymer alloys based on a 2-D FEA of a cylindrical inclusion embedded in a thin plate under plane-stress condition omitting the inter-particle interaction. However, these analyses seem not to be directly applicable to the case of spherical particles embedded in a thick specimen under triaxial stress or indeed to the crack-particle interaction problem.

In the present work, a three-dimensional elasto-plastic finite element analysis is carried out to study the toughening processes in rigid-rigid polymer blends with initially spherical secondary-phase plastic particles in a periodic face-centred cubic layout accounting for the particle-particle interaction by a face-centred cuboidal cell model. In future work, we intend to incorporate appropriate criteria for cavitation and /or debonding in the 3-D FEA in order to quantify these toughening processes. Howver, independent experiments need to be designed to measure the cavitational and/or debonding strengths for a given rigid-rigid polymer blend as these properties are generally not known.

MICROMECHANICAL MODEL

FCC Cell Model

The representative unit of the periodic FCC microstructure is shown in Figure 1. Under symmetric loading, only one-eighth of the face-centred cuboidal cell is needed for numerical analysis due to the periodic symmetry of the problem.

Periodic Boundary Conditions

To satisfy the requirements of periodic symmetry, the following constraint equations on the corresponding surfaces of the cell are imposed:

$$u_z = U_{z0}, \qquad t_x = t_y = 0 \qquad \text{at } z = a_0$$
$$u_x = U_{x0}, \qquad t_y = t_z = 0 \qquad \text{at } x = a_0 \qquad (1)$$
$$u_y = U_{y0}, \qquad t_x = t_z = 0 \qquad \text{at } y = a_0$$

and

$$u_z = 0, \qquad t_x = t_y = 0 \qquad \text{at } z = 0$$
$$u_x = 0, \qquad t_y = t_z = 0 \qquad \text{at } x = 0 \qquad (2)$$
$$u_y = 0, \qquad t_x = t_z = 0 \qquad \text{at } y = 0$$

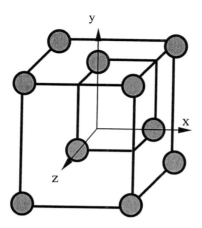

Figure 1. Face-centred cuboidal cell model. Reproduced with permission from reference 6. Kluwer Academic Publishers, 1998

where a_0 is the cell length, u_x, u_y and u_z are components of the displacement vector $\mathbf{u}$ along the x-, y- and z- directions, t_x, t_y and t_z are the x-, y- and z-components of the surface traction vector $\mathbf{t}$, U_{x0}, U_{y0} and U_{z0} are displacement constants. The surfaces of $x = 0$, $y = 0$ and $z = 0$ are symmetric planes, while the surfaces of $x = a_0$, $y = a_0$ and $z = a_0$ are kept parallel with respect to their original shapes during deformation to satisfy the periodic symmetry requirements.

Definition of Effective Stress and Strain

The effective stress σ^e and strain ε^e are obtained by averaging the local stress σ and strain ε in the face-centred cuboidal cell, that is,

$$\sigma^e = \frac{1}{V_\Omega} \int_{V_\Omega} \sigma dV = \frac{1}{V_\Omega} \int_{S_\Omega} \mathbf{r} \otimes \mathbf{t} dS \qquad (3)$$

$$\varepsilon^e = \frac{1}{V_\Omega} \int_{V_\Omega} \varepsilon dV = \varepsilon^0 \qquad (4)$$

where $\mathbf{r}$ is the position vector and $\mathbf{t}$ is the traction vector on the cell surface, ε^0 is the constant strain tensor dependent on the constant normal displacement on the cell surface, V_Ω and S_Ω represent the cell volume and surface, respectively. We use the equilibrium condition without a volume force, i.e. $\nabla \cdot \sigma = 0$, to obtain the last equality of eq 3.

FINITE ELEMENT SIMULATION

Stress Distributions

3-D elasto-plastic finite element analysis was carried out using the ABAQUS program on an ALPHA STATION 500. We study the PEI/PC system with PC as second-phase particles and PEI as matrix, respectively. Both materials are governed by the Von Mises criterion with piece-wise linear stress-strain relations given by Sue, Pearson & Yee (9). PC has a Young's modulus of 2400 MPa and a Poisson's ratio of 0.42. PEI has a Young's modulus of 3400 MPa and a Poisson's ratio of 0.40. The ratio of particle diameter to cell length is chosen as 0.5, 0.4 and 0.2, corresponding to spacing ratio of 0.5, 0.6 and 0.8. The volume fractions (f_i) of PC particles are hence 26.18%, 13.40% and 1.68%. The PATRAN program was used to generate the mesh automatically in the one-eighth face-centred cuboidal cell model. For comparison, we use the same mesh for the PEI/void system with PC particles replaced by voids. The applied stress system $T_x : T_y : T_z = 0 : 0 : 1$ represents macroscopic uniaxial tension; and $T_x : T_y : T_z = 1.5 : 2.0 : 2.5$ corresponds to a high tensile triaxial stress state ahead of the elastic-plastic boundary of the crack tip region (11).

The contour plots of Von Mises stress, hydrostatic pressure and direct stress in PEI/PC under the two applied stress systems are shown in Figures 2 to 4 for an applied z-axial strain of 0.02. We can see that the stresses inside PC particles are non-uniform, which is different from those inside rubber particles (9,10). Maximum Von Mises stress, dilatational stress and direct stress in the PEI matrix are all reached in the equatorial regime of the PC particles. Local dilatational stress increases while local Von Mises stress decreases with increasing stress triaxiality, which shows that a high tensile triaxial stress state is beneficial for debonding, cavitation and crazing whereas it defers the development of shear deformation. Hence, the plane-strain triaxial stress condition in front of the crack tip has to be relieved in order to activate the shear yielding mechanism for a thick specimen with a macro-crack.

Stress Concentrations

The maximum Von Mises stress concentration in the matrix for the PEI/PC blends at three different particle volume fractions of 1.68%, 13.40% and 26.18% under uniaxial tension is shown in Figure 5. It is seen that the shear stress concentration increases with increasing PC particle volume fraction. Two distinct peaks occur as the PC particles begin to undergo anelastic and plastic deformations, the spacing of which decreases with increasing PC particle volume fraction. The maximum shear stress concentration is constant at 1.15 at the elastic stage and rises to 1.18 as anelastic deformation occurs in the PC phase for the PEI/PC system at a low f_i of 1.68%. Sue, Pearson and Yee reported a shear stress concentration factor of ~1.2 in the elastic regime and about 1.3 as the PC inclusion begins to undergo anelastic deformation by 2-D FE simulation of a cylindrical inclusion embedded in a thin plate

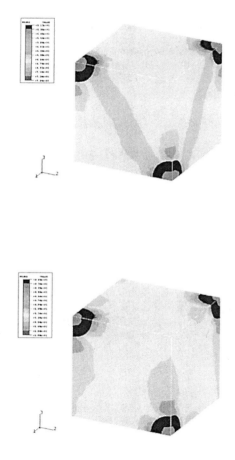

Figure 2. Contour plots of Von Mises stress in PEI/PC blends for two applied stress

systems: (a) $T_x : T_y : T_z = 0 : 0 : 1$, *(b)* $T_x : T_y : T_z = 1.5 : 2.0 : 2.5$.

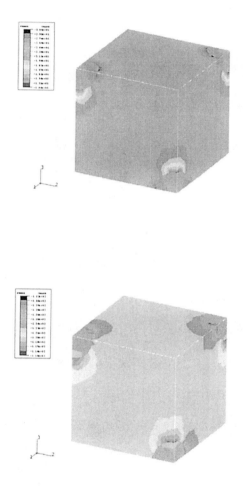

Figure 3. Contour plots of hydrostatic pressure in PEI/PC blends for two applied stress systems: (a) $T_x : T_y : T_z = 0 : 0 : 1$, (b) $T_x : T_y : T_z = 1.5 : 2.0 : 2.5$.

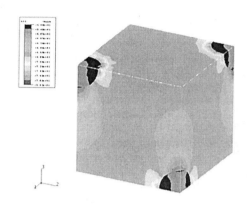

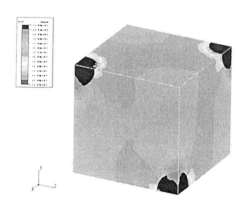

Figure 4. Contour plots of direct stress in PEI/PC blends for two applied stress systems: (a) $T_x : T_y : T_z = 0 : 0 : 1$, (b) $T_x : T_y : T_z = 1.5 : 2.0 : 2.5$.

under uniaxial tension (9). The 2-D results by Sue *et. al.* agree qualitatively with our 3-D results in the prediction of the general trend of the effects of anelastic and plastic deformation of PC particles to initiate shear yielding. Localised plane strain shear bands can be formed under uniaxial tension even without the cavitation process because the shear stress concentration factor has reached 1.15.

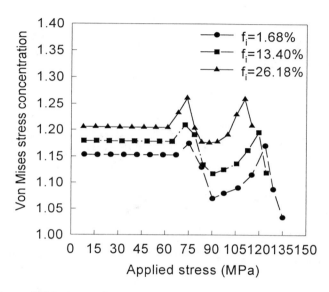

Figure 5. Maximum shear stress concentration versus applied stress for PEI/PC blends at three different particle volume fractions under uniaxial tension.

Figure 6 shows the maximum shear stress, dilatational stress and direct stress concentrations for both PEI/PC and PEI/void systems at a particle volume fraction of 1.68% under uniaxial tension. Clearly all these stresses are elevated when the particles are replaced by the voids. The maximum shear stress concentration is as high as 1.91 for the PEI/void system in contrast to 1.18 for the PEI/PC system at the same f_i.

Effective Stress-Strain Relation

The effective elasto-plastic stress-strain relations for the PEI/PC blends with f_i equal to 1.68%, 13.40% and 26.18% under uniaxial tension are shown in Figure 7. Likewise, the effective elasto-plastic stress-strain relations for both PEI/PC and PEI/void systems with f_i = 1.68% under uniaxial tension and the triaxial stress system ahead of the crack tip elastic-plastic boundary are shown in Figure 8.

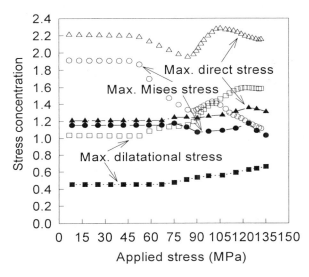

Figure 6. Maximum shear stress, dilatational stress and direct stress concentration versus applied stress for PEI/PC and PEI/void systems at a particle volume fraction of 1.68% under uniaxial tension. (Solid symbols: PEI/PC system. Open symbols: PEI/void system).

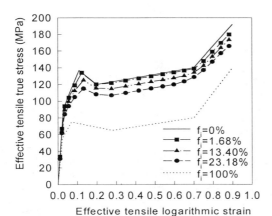

Figure 7. Effective elasto-plastic stress-strain relations for PEI/PC blends at three different particle volume fractions under uniaxial tension.

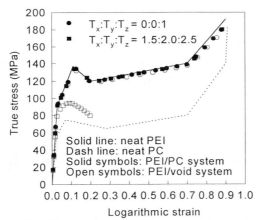

Figure 8. Effective elasto-plastic stress-strain relations for PEI/PC and PEI/void systems at a particle volume fraction of 1.68% under uniaxial tension and imposed triaxial stress in front of a crack tip.

It can be seen from Figure 7 that the effective stress-strain relations for the PEI/PC system are similar to the neat PEI and PC with a strain-hardening effect. The effective stress decreases with increasing PC particle volume fraction. The yield behaviour of the PEI/void system is quite different to that of the PEI/PC system with a much lower effective shear yield stress under the high tensile triaxial stress associated with a crack tip although they are similar under uniaxial tension, Figure 8. Unstable void growth may occur in the PEI/void system at high stress triaxiality as the macroscopic logarithmic strain reaches around 20%.

CONCLUSIONS

The effective stress level decreases with increasing PC particle volume fraction. The PEI/void system has a much lower effective yield stress under high tensile triaxial stress associated with a crack tip than the PEI/PC system although they have similar effective yield stresses under uniaxial tension. The local stress concentration increases with increasing PC particle volume fraction. Two distinct peaks occur in the curve of maximum shear stress concentration versus applied stress as PC particles begin to undergo anelastic and plastic deformations. The local dilatational stress concentration increases while the local Von Mises stress concentration decreases with increasing stress triaxiality. These results confirm that a high tensile triaxial stress state is beneficial for debonding, cavitation and crazing, whereas it defers the development of shear deformation. Localised plane-strain shear bands can form under uniaxial tension without resorting to the debonding-cavitation process; whereas the triaxial stress plane-strain condition associated with a macro-crack in a thick specimen has to be relieved so that extensive plastic deformation can be developed to produce toughening. Partial debonding-cavitation at the PEI/PC interface can promote shear yielding while still keeping the craze-stabilising and crack-bridging functions of the PC domains.

ACKNOWLEDGEMENTS

The authors wish to thank the Australian Research Council (ARC) for the continuing support of the polymer blends project. X.-H. Chen is grateful for the ARC Postdoctoral Fellowship.

REFERENCES

1. Bucknall, C. B. *Toughened Plastics*, Applied Science Publishers LTD, London, 1977.
2. Riew, C. K.; Kinloch, A. J. *Toughened Plastics II*, American Chemical Society, Washington, DC, 1996.
3. Yee, A. F.; Pearson, R. A. *J. Mater. Sci.* **1986**, *21*, 2462.
4. Chen, X. H.; Mai, Y.-W. *Key Engineering Materials* **1998**, *137*, 115.
5. Chen, X. H.; Mai, Y.-W. *Key Engineering Materials* **1998**, *145-149*, 233.
6. Chen, X. H.; Mai, Y.-W. *J. Mater. Sci.* **1998**, 33, 3529.
7. Chen, X. H.; Mai, Y.-W. *J. Mater. Sci.* **1999**, 34, 2139.
8. Chen, X. H.; Mai, Y.-W. *Polymer Engineering and Science* **1998**, 38, 1763.
9. Sue, H. J.; Pearson, R. A.; Yee, A. F. *Polymer Engineering and Science* **1991**, *31*, 793.
10. Wu, J. S.; Mai, Y. -W.; Yee, A. F. *J. Mater. Sci.* **1994**, *29*, 4510.
11. Knott, J. F. *Fundamental of Fracture Mechanics*, John Wiley & Sons, New York, 1976.

EXPERIMENTAL STUDIES

Chapter 7

Novel Mechanisms of Toughening
Semi-Crystalline Polymers

A. S. Argon, Z. Bartczak[1], R. E. Cohen, and O. K. Muratoglu[2]

Massachusetts Institute of Technology, Cambridge, MA 02139

Many normally ductile semi-crystalline polymers such as polyamide-6 (or 66) (nylon), high density polyethylene, (HDPE), and isotactic polypropylene (iPP) are known to be brittle under impact loading.

Here we present some general principles for improvement of toughness that rely on the reduction of plastic resistance of the cavitated matrix. As typical cases, we discuss the mechanisms of the dramatic toughness jumps achievable in both nylon and HDPE through the incorporation of either flexible rubbery particles or rigid $CaCO_3$ particles. In both cases these result in the establishment of a material component of reduced plastic resistance in the form of a layer of oriented crystallization of well defined thickness around the particles. The toughness jumps occur when the average inter-particle ligament thickness is reduced below a critical value, specific to the particular polymer, and the component of reduced plastic resistance percolates through the structure.

Introduction

Many polymers are capable of exhibiting mechanical behavior ranging from very tough to quite brittle. There have been a large number of ad-hoc approaches administered to remedy the brittle behavior of polymers, which have been based on empiricism, promoting much *ex-post-facto* rationalization resulting in a rich collection of prescriptions for incorporation of particles of certain size or spacing and to accomplish a variety of processing steps to control morphology. Some successful

[1]Permanent address: Polish Academy of Sciences, Center of Molecular and Macromolecular Studies, 90–363 Lódz, Poland.

[2]Present affiliation: Massachusetts General Hospital, Orthopedic Biomechanics Laboratory, Boston MA 02114.

and relatively well understood principles applicable to glassy polymers such as the coaxing of cavitational plasticity through crazing have been considered also effective for semi-crystalline polymers - often without any mechanistic evidence or rationale. On the whole, there is a widespread expectation in the "culture" of polymer research that beneficial properties of one polymer can be incorporated into another with deficient properties by blending. Such exercises frequently had remarkable results based on mechanisms that have been clarified much later by careful detective work and modeling.

Here we start by recalling some elementary principles that control brittle or tough behavior in intrinsically brittle solids which have well established fracture transitions. We then demonstrate, through some polymer systems and particularly through the behavior of rubber toughenable polyamide 6 or 66 (Nylon-6 or 66) and similarly modified polyethylene how these principles apply to polymers. These will serve as clear examples of mechanism based modifications of behavior having wide ranging applicability.

Sources Of Brittleness And Mechanisms Of Toughening

Brittle to Ductile Transitions

Experiments indicate that many semi-crystalline polymers such as HDPE and the various types of nylon, are all intrinsically brittle, in the sense that crystalline solids have been classified as either intrinsically brittle or ductile (1,2). Such solids have a brittle to ductile fracture transition that manifests itself in brittle behavior at low temperatures and/or high strain rates where a crack can propagate with little resistance. In their brittle range the critical stress intensity factors, K_{IC} of such polymers are in the range of 0.5 Mpa $\sqrt{m}$.

In the corresponding problem the brittle to ductile (BD) transition in steels it has been useful to think in terms of a relatively temperature independent and flaw-governed brittle strength, σ_B, being in competition with energy absorbing plastic behavior that is characterized by a yield strength Y_o having considerable temperature and strain rate dependence. Figure 1a illustrates this behavior schematically in terms of the well known Davidenkov diagram (3) which shows that for a given strain rate there should be a transition from brittle to ductile behavior at a certain temperature, and that this transition temperature increases with increasing strain rate. Moreover, while the brittle strength relates to a response governed by a principal tensile stress, the yield behavior responds only to a critical level of the effective (deviatoric) stress, σ_e. In the presence of sharp notches, individual normal stress components can be substantially augmented by a negative pressure, while the effective stress remains equal to Y_O in

producing plastic flow. This will result in a marked increase of the brittle to ductile transition temperature, as is well known in the notch impact testing of structural alloys. In polymer technology these well known phenomena often become obscured

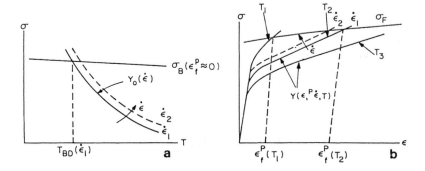

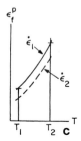

Figure 1 Schematic representation of brittle to ductile transition in fracture: a) the "Davidenkov" diagram depicting BD transition with increasing temperature and decreasing strain rate; b and c) decrease of strain to fracture with decreasing temperature and increasing strain hardening rate resulting from increases of plastic resistance.

by morphological detail and are not recognized in dealing with the brittleness/toughness transitions.

As stated above, experiments indicate that the driving force for brittle crack propagation in many intrinsically brittle polymers, including semi-crystalline polymers, such as e.g., Nylon-6 or 66, is a K_{IC} of the order of 0.5MPa√m. When structural imperfections such as poorly adhered inorganic inclusions, deliberately incorporated particles with no grafting, cavities, and the like, are present in a size range of 25-50 μm, they will result in a brittle strength in the range of 40-55 MPa.

When structural imperfections are well controlled and are only in the micron range, the brittle strength of polymers is well above the initial yield strength, and as a consequence they will deform plastically. Once plastic response is initiated it will result in neutralization of some of the effects of the imperfections and eventually result in molecular alignment or texture development (4,5) that can significantly elevate the fracture toughness across the extension direction.

As Figure 1b depicts, however, strain hardening usually increases the plastic deformation resistance faster than increasing the tensile strength, resulting in eventual fracture. Here too, the strain induced elevation of fracture resistance is relatively independent of strain rate or temperature, while the rising plastic resistance $Y(\varepsilon^p)$ is more strongly temperature and strain rate dependent. Thus, as for the initial response of the polymer represented in Figure 1a, the final fracture strain decreases with decreasing temperature and increasing strain rate as depicted in Figure 1c. Clearly, the processes involved in the strain hardening and the associated improvement of fracture strength can be complex and different in different polymers and offer possibilities in morphology control. Nevertheless, the basic response characteristics demonstrated in Figure 1 are generic to intrically brittle solids, and apply to polymers as well.

Role of Particle Modification in Toughening

The subject of interest here is the role of particles that are frequently incorporated into polymers to counteract brittleness. We will show that the effect of all particle modification of polymers can be understood readily in the light of the simple principles illustrated in Figure 1.

In what follows, we concentrate on the cases of particle incorporation into semi-crystalline polymers where the behavior pattern includes decohesion or cavitation of particles, that, in the most effective cases, promotes substantial subsequent plastic response. Very similar phenomenology of toughening also applies to thermosets which we have discussed in detail elsewhere (6,7). We exclude from our discussion the generally well understood cases of compliant particles that promote conventional craze plasticity in flexible-chain glassy polymers (8-11); the process of plasticity by planar zone cavitation of rubbery block domains (8, 12-14) and the very attractive cases of pre-packaged plasticizing diluents, accelerating craze plasticity (15-20).

In many epoxies (21) and semi-crystalline polymers (22,23) the incorporation of both rubbery or stiff inorganic particles of sub-micron to micron dimensions have resulted in considerable toughening of either brittle polymers or reasonably ductile polymers with notch brittleness behavior. In nearly all such instances the pre-requisite to toughening has been decohesion or internal cavitation of the particles. It has often been emphasized that for rubbery particles grafting is essential. Since, however, blending of rubber in ungrafted form usually results in a very wide particle size distribution often ranging well above 50μm, while grafting of particles results in a far tighter distribution and considerably smaller range of sizes well under 10μm, for the same volume fraction, it is not at all clear that grafting which prevents decohesion is essential. Particularly when in other polymers decohesion of inorganic particles of a narrow size distribution accomplishes the same end of toughening.

For the purpose of illustration consider as a generic example of a polymer with a brittle strength σ_B governed by some extrinsic imperfections resulting in a fracture toughness K_{ICo} or corresponding critical energy release rate G_{ICo}, containing a set of rubbery particles of volume fraction c, and of subcritical dimensions, when compared to the dimensions of the imperfections that control σ_B. We now consider two possible loading histories and behavior patterns.

To demonstrate these points Argon et al. (7) have discussed two likely limiting cases involving scenarios in which a) the initial yield strength Y_o exceeds σ_B and; the incorporated particles of volume fraction c cavitate during elastic loading but this does not result in general yield, and b) in the same scenario of above, upon particle cavitation, general yielding occurs under the reduced deviatoric yield strength of $(1-c)$ Y_o. In the first case the only benefitial effect of the cavitation is a modest effect of crack tip shielding (7). In the second scenario cavitation of particles results in substantial toughening of the polymer in a manner explored by a number of previous investigators (21,24).

The most remarkable example of the above scenario of interest here is the notch sensitivity of rubber modified nylon, investigated previously by Wu (22,23) and Borggreve et al.(25). Wu has noted that the notch brittleness of unmodified homo-nylon is improved dramatically upon the incorporation of a finely dispersed, grafted rubber phase by blending. He noted, however, that the toughening effect did not correlate directly with either the rubber particle size or with its volume fraction, but rather with the inter-particle ligament dimension as is shown in Figure 2. For the rubber modified nylon system the material exhibits notch brittleness at room temperature for ligament dimensions greater than about 0.3μm, but for dimensions less than this, it exhibits a dramatic toughness jump. Increasing volume fractions of incorporated rubber increases the impact energy further in a predictable manner, provided the critical ligament dimension is maintained. Explanations given by Wu and others based on field theory arguments such as stress field overlap, local state of either plane stress or plane strain, etc., can not be taken seriously since field theory is scale independent. Clearly, this dramatic effect that scales with a physical dimension must be a consequence of a specific material constitution at this particular threshold dimension. In what follows, we demonstrate that this is indeed the case, and that the critical ligament dimension is a consequence of a specific morphology of crystallization of the nylon matrix in between the rubber particles. Moreover, once

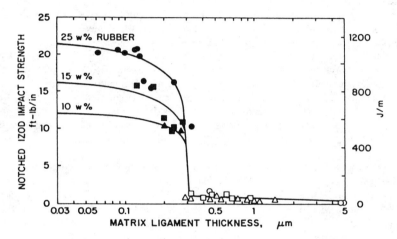

Figure 2 Jump in Izod impact energy of rubber modified Nylon-66 when the matrix ligament thickness becomes less than a critical dimension, independent of volume fraction or particle size (Reprinted with permission from reference 23, Copyright 1988)

understood in this manner, the mechanism has widespread applicability to other semi-crystalline systems with equally dramatic results, as we demonstrate for HDPE.

Toughening Of Semi-Crystalline Nylon By Rubber Modification.

Morphology of Semi-crystalline Nylon-6 and Deformation Mechanisms

Nylon is a semi-crystalline polymer in which the crystalline component at a volume fraction of roughly 0.5 is in the form of lamellae of c.a. 12nm thickness, having a monoclinic crystal structure as depicted in Figure 3a, with two prominent chain slip systems: (001)[010] and (100)[010]. In the usual melt-grown form the morphology of the material is spherulitic (26) in which the crystalline lamellae, are separated by alternating layers of amorphous material. The plasticity of this material and how it can acquire a preferred deformation texture with increasing plastic strain was studied in detail (5), providing information on deformation resistances of the principal slip systems of crystalline nylon and have established that the (001)[010] chain slip system, which contains hydrogen bonds in its plane, has the lowest slip resistance at 16.2MPa at room temperature, while the second best chain slip system, (100)[010], and the transverse system, (001)[100], have slip resistances of 23.2MPa, respectively.

To explain the unusual ligament thickness correlation of the toughness jump noted by Wu, Muratoglu et al., (27) postulated and subsequently demonstrated, that in the preparation of the rubber modified-Nylon-66, the rubber (ethylene-propylene- diene rubber), grafted with maleic anhydride, EPDR-g-MA undergoes phase separation first, and the nylon crystallizes subsequently in a preferential manner with the low energy/low plastic shear resistance (001) crystallographic planes becoming parallel to the rubber/nylon interfaces. This results in a specially oriented crystalline environment of a characteristic thickness around every rubbery particle. In the tough material with ligament thickness less than 0.3μm the specially oriented crystalline material layers of homo-nylon, of adjacent rubbery particles, touch and such oriented material percolates through the entire structure, as depicted in Figure 3b. This gives rise to a morphology of lowered overall plastic resistance.

When the inter-particle ligament thickness exceeds 0.3μm the specially oriented crystalline material layers become separated and no longer percolate through the structure, as much of the background material becomes populated with a polycrystalline aggregate of randomly oriented lamellae as depicted in Figure 3c. This results in a morphology of significantly higher overall plastic resistance. In this postulated behavior pattern, deformation of the rubber-modified-nylon, in its tough behavior begins first by cavitation of the rubbery domains, ending up with a randomly arranged cellular material in which the cell walls are composed of preferentially arranged lamellar crystallites with the best slip planes being arranged parallel to the interfaces that have now been converted to free surfaces as depicted in Figure 4a.

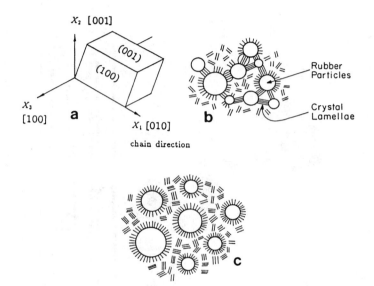

Figure 3 a) Crystallography of monoclinic polyamide-6 (Nylon-6); Schematic representation of the layer of crystallites of preferred orientations: b) when ligament thickness is less than twice the thickness of the oriented crystallization layer around the particles and material with reduced plastic resistance percolates through the blend; c) when ligament thickness is larger than twice the oriented crystallization layer and material exhibits brittle behavior.

106

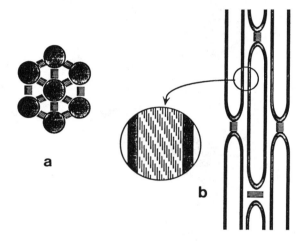

Figure 4 Diagram showing stages of the tough response of rubber modified Nylon-6: a) initial morphology of particles and thin ligaments containing only the oriented crystallization layer; and b) the cavitated and stretched out matrix showing elongated sausage shaped cavities.

Under these conditions the sectors of the cavity surfaces parallel and perpendicular to the principal stretch direction will deform little but the sectors making roughly 45° angles with the extension direction will undergo deformation readily by chain slip (or by transverse slip) by virtue of the ideal orientation of the lamellar material in these sectors for such slip response. This should then result in the stretching-out of the ligaments to form sausage shaped cavities as depicted in Figure 4b.

These postulated structures and stages of deformation have been verified in considerable detail by various forms of microscopy. Figure 5a shows a typical transmission electron micrograph of several rubbery particles and the morphology of the surrounding semi-crystalline nylon in which the dark regions, stained by phospho-tungstic acid are the amorphous fractions of the homo-nylon and the stringy, light, regions are the ribbon shaped crystallites (26). In these crystallites the readily shearable (001) planes are perpendicular to the crystallites i.e. parallel to the rubber/nylon interface as hypothesized. Figure 5b shows regions of the cavitated and stretched-out material that had undergone large strain deformation.

While the evidence shown in Figures 5a and 5b in support of the hypothesis of the special toughening mechanism is attractive, it falls short of being fully convincing since it is difficult to demonstrate the hypothesized alignment of the (001) planes parallel to the interfaces by any definitive scattering technique and the anisotropic plastic resistance. Therefore, the hypothesis was tested in separate experiments in which films of Nylon-6, in thickness ranging from 0.15-2.15μm were prepared, sandwiched between two thin layers of EPDR rubber to create incoherent planar interfaces similar to those occurring in the rubber-particle-modified-nylon, against which the nylon film can crystallize. Such films were removed and allowed to undergo unconstrained stress relief and structural relaxation in an elaborate preparation technique discussed elsewhere (28). Transmission electron microscopy and wide angle x-ray diffraction from the (001) planes of the thinnest films indicated that nearly all crystalline lamellae in these films were perpendicular to the free surfaces of the films as expected, and that moreover, the orientation distribution of the (001) planes were predominantly parallel to the surfaces in the thinnest film as shown in Figure 6a. As the figure shows, this distribution became progressively more diffuse with increasing film thickness - evidently due to the random orientation of lamellar domains in the center regions of the thick films. The more direct verification of the correctness of the hypothesis was obtained from the stress-strain curves of these free-standing thin films, shown in Figure 6b. Clearly, in the thinnest film the best slip systems are all parallel to the free surfaces and can not take part in the deformation, leaving only the next best chain slip systems and transverse systems available with plastic shear resistances of c.a. 23 MPa to provide the extensional strain. Thus, the flow stress of the 0.15μm thick films, at 110 MPa is taken as the polycrystal aggregate average deformation resistance of these systems with a definite fraction of crystals, with their chain axes parallel to the extension direction, acting as rigid inclusions (29,30). As the films become thicker and a substantial fraction of the best (001)[010] slip systems become randomly oriented in the interior portions of the film, the aggregate flow stress drops systematically, until for a film of 2.15μm, it reaches a

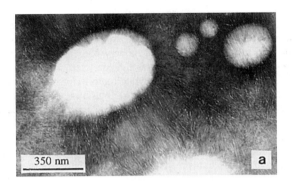

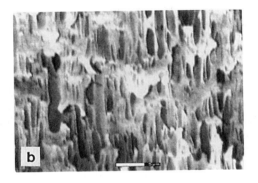

Figure 5 a) TEM micrograph showing the oriented lamellar (ribbon shaped) crystals around the rubbery particles in Nylon-6. The slip planes of the best slip system (001)[010] are perpendicular to the ribbon shaped lamellae. The dark regions represent the amorphous component stained by phospho-tungstic acid. b) SEM micrograph of stretched out sausage shaped cavities in Nylon-6 containing 0.19 vol fraction of EPDR rubber, after a large strain tensile deformation excursion (Reprinted with permission from reference 43, Copyright 1995).

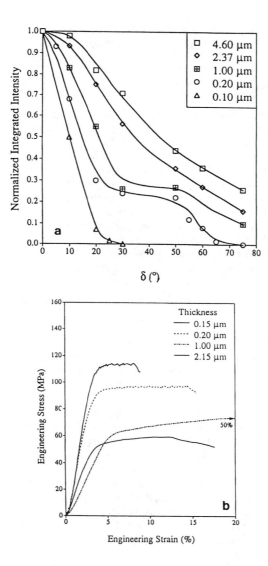

Figure 6 a) Normalized distribution of diffracted x-ray intensity from the (001) planes of thin films of Nylon-6, as a function of tilt angle of the plane of the x-ray beam and the normal of the surface of the thin films; b) stress strain curves of free-standing thin films of Nylon-6 similar to those of part (a), but crystallized between thin polystyrene layers (Reprinted with permission from reference 43, Copyright 1995).

level of 55MPa, which must now be considered as the polycrystal aggregate average deformation resistance, over all orientations - including some lamellae acting as rigid inclusions.

Toughness of Rubber-Modified Nylon

Once the critical ligament thickness condition in the EPDR-rubber-modified-nylon has been satisfied to reduce the deformation resistance and thereby rectify the notch brittleness, the cavitated material was found to become very tough indeed.

To make a meaningful determination of the tough behavior of such rubber-modified-material, special double cantilever beam specimens prepared of a Nylon-66 blend with 20% EPDR rubber, were used to measure the crack growth resistance (31). Figure 7 shows the load/pin-displacement curve in such an experiment. In the rising portion of the curve the arms of the specimen are pried open statically, and the crack is slowly extended, until at the load maximum a steady state crack opening angle of close to 47^0, is reached, where the crack begins to grow stably under decreasing pin-load.

To obtain a measure of the material toughness the information in Figure 7 was used to determine the crack driving force J, where the peak value of the load was taken to correspond to the steady state value J_{ss}.. Using the usual means of evaluation of the J integral (32) based on the given specimen dimensions and material information reported by Lin and Argon (33), a steady state crack driving force of J_{ss} = 754 J/m^2 was obtained together with a tearing modulus of, T = 39.5 based on the steady state plastic crack opening angle of 47^0. Tearing moduli of this magnitude are comparable with those reported for some of the toughest pressure vessel steels (34). If it is recognized that nylon has a density of only about 1/7th of that of steel, the toughness becomes even more impressive.

In contrast with this dramatically tough behavior is the extremely brittle behavior of an ungrafted blend with the same concentration of rubber but with very non-uniform particle size distribution and with some particle sizes exceeding 50μm, acting as the strength controlling flaws and resulting in an Izod energy absorption of a nearly negligible amount of 69 J/m. In this case the ligament dimensions were in excess of 10μm, as is clear from the appearance of the fracture surface shown in Figure 8, and far above the level of what would be required for a tough matrix. Clearly, the fact that the rubber was not grafted in this instance is far less important than the consequence of the attendant large thickness of the average dimensions of ligaments.

The special toughening mechanism relying on the oriented crystalline layers in the ligaments at or under the critical thickness dimension Λ_c, has been explored by Tzika et al. (35) by a finite element study using anisotropic plasticity to represent the specific plastic resistance of these layers. In this study the authors demonstrate the initiation of plastic response at roughly the expected angle of 45° between the poles and the equator of the cavities replacing the particles – as discussed in connection with Figure 4a, and the eventual stretching out of the initially spherical shaped cavities into sausage shaped entities as shown in Figure 5b. Figure 9 shows a typical sequence of

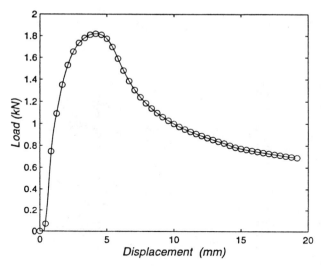

Figure 7 *Load/pin-displacement curve of an elongated "compact" tension specimen of 5.4 mm thickness, of a rubber modified Nylon-6 sample, using stable crack extension with a steady state plastic crack opening angle of (Reprinted with permission from reference 43, Copyright 1995).*

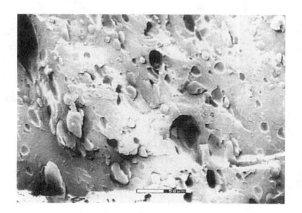

Figure 8 *Fracture surface of notch brittle Nylon-6 with a rubber particle component with 0.19 volume fraction with no grafting and no particle size control where the cavities that result from premature decohesion act as supercritical flaws to give brittle behavior (Reprinted with permission from reference 43, Copyright 1995).*

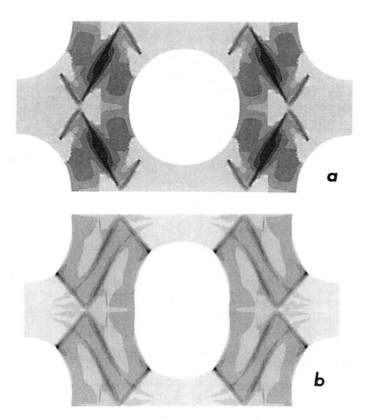

Figure 9 Finite element modeling of the sausage shaped elongations by anisotropic plastic flow of a typical cavity environment in nylon: a) at just fully developed plastic flow, b) at extension ratio of 1.25.   (Reproduced with permission from reference 35, to be published).

development of this process of cavity stretching in the direction of principal extension.

Toughening Of Semi-Crystalline High Density Polyethylene With Both Rubber And Calcium Carbonate Particles

The principle of toughening based on the preferentially oriented crystalline layer around particles that was introduced above for Nylon-6 and 66, has been explored fully also in HDPE which too has a well known notch brittleness problem under impact conditions. Since the toughening mechanism relies only on preferential crystallization of the matrix homopolymer on incoherent interfaces between the homopolymer and the incorporated inclusion particles, the process in HDPE was studied in systems in which the incorporated particles were alternatively either EPDM rubber, as was the case in Nylon blends, or equiaxed $CaCO_3$. In both cases the volume fractions of particles were 0.22. Fu et al. (36), reported critical ligament dimensions for the toughness jump in HDPE to be well above 1.0μm.

To establish a proper mechanistic comparison with the nylon system, experiments were carried out where HDPE films of 4 different thicknesses ranging from 0.1μm upto 1.2μm (0.1, 0.2, 0.6, 1.2) were crystallized between two flat planar rubber layers or alternatively formed on the (104) plane surfaces of large calcite crystals, followed by a determination of the orientation distribution of the HDPE lamellar crystals by x-ray diffraction pole figures to obtain information comparable to that presented for Nylon-6 in Figure 6. Here we note first that the crystal structure of PE is orthorhombic where the critical resolved shear stresses at room temperature for the three prominent chain slip systems are: 7.2MPa for (100) [001], 15.6MPa for (010)[001] and in excess of 13MPa for (110)[001], and 12.2 MPa for the principal transverse slip system of (100)[010] (37). The normalized diffracted intensities from the (200) and (110) planes of the PE crystals in these thin films are shown in Figures 10 and 11 as a function of the angle of inclination of the planes measured away from the surface normals of the films. The figures show that for the film with thicknesses at or, under 0.6μm there is a distinct clustering of the (100) planes parallel to the surfaces of the film which positions the (110) planes at roughly a 62° inclination relative to the film surfaces, regardless of whether they crystallized against rubber or $CaCO_3$. Thus, we conclude first, that for PE, as was in the case of nylon, there is a very distinct preference for the low-energy/low-shear-resistance crystallographic planes to orient themselves parallel to incoherent internal interfaces during crystallization. Such oriented crystallization of PE lamellae against $CaCO_3$ surfaces had been noted earlier by Chacko et al. (38), in a TEM study where, however, the consequences of such oriented crystallization had not been explored. Moreover, results of the type shown in Figures 10 and 11 indicate that the thickness of the oriented crystalline layer is of the order of 0.6μm, verifying the possibility that the critical ligament thickness for the toughness jump is likely to be close to the value reported by Fu, et al. (36). Of greater significance to us here is the fact that nearly identical results were obtainable for both PE/EPDM rubber interfaces as well as for PE/$CaCO_3$ interfaces,

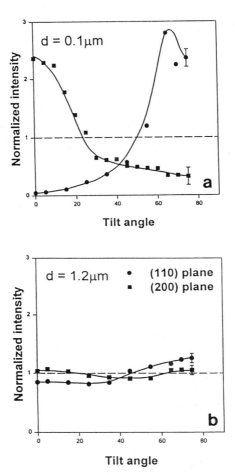

Figure 10 Orientation distribution of normals of diffracting planes (200) (■) and (110) (●) relative to normal of film surface in thin PE films crystallized between two rubber layers: a) film of 0.1 μm thickness, b) film of 1.2 μm thickness.

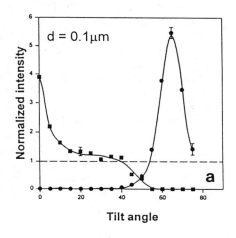

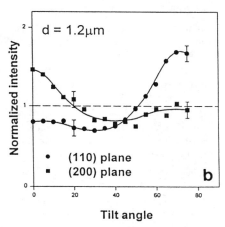

Figure 11 Orientation distribution of normals of diffracting planes (200) (■) and (110) (●) relative to normal of film surface in thin PE films cast against a (104) crystallographic plane of calcite crystal: a) film of 0.1 µm thickness, b) film of 1.2 µm thickness.

suggesting that the toughening effect obtainable with both of these particles of radically different elastic properties should be the same and that the notion of percolation of material with low plastic resistance is the key controlling factor. That this is indeed the case for $CaCO_3$ particles in HDPE is shown in Figure 12 demonstrating a toughness jump at roughly about 0.8μm ligament thickness at room temperature.

Almost identical results were obtained also for the case of EPDM rubber particles at room temperature. In both cases the tough behavior below the critical ligament thickness involves either particle cavitation (as in the case of EPDM) or particle decohesion (as in the case of $CaCO_3$) which then permits the development of the large energy absorbing anisotropic plastic flow process in the freed ligaments resulting in the dramatic forms of cavity elongation. That this is the case is shown in Figure 13a for EPDM rubber particles and 13b for $CaCO_3$ particles in cryo-fracture-produced sections of the fractured Izod specimens along planes perpendicular to the fracture surface and parallel to the surfaces of the Izod specimens. The specific inclination of the elongated cavities relative to the fracture surfaces is a result of the large anisotropic plastic flow in the flank regions of the crack upon the immediate passage of the propagating crack (39). The important role of prior cavitation in the development of the full toughening effect is demonstrated in Figure 14 which shows the temperature dependence of the Izod fracture energy for both the rubber, (ethylene octane rubber EOR) and $CaCO_3$ particle-modified-blends for ligament thicknesses less than the threshold dimension of 0.8μm. In the case EOR rubber 4 types were used with glass transition temperatures of 3 types differing roughly by 15°C (EOR-4 with Tg ≅ -15C; EOR-1,2,3 with Tg ≅ -30C). Blends with all types of rubber particles exhibit impressive Izod toughness energies above -10C where all particles behave in a compliant rubbery manner to cavitate internally at the start of deformation to allow plastic cavity elongation. Below -15C for EOR-4 rubber and below -30C for the three other EOR-1,2,3 rubbers the grafted and well adhered particles with now negligible elastic property mismatch in their glassy state relative to HDPE behave as rigid inclusions and do not cavitate internally and can not debond. The resulting fracture energy is disappointingly low at around 50J/m. In comparison, blends with $CaCO_3$ particles of 0.7μm diameter, the fracture energy remains high and relatively temperature independent since in all cases the particles decohere and permit the process of plastic cavity elongation to fully develop. Interestingly, the figure shows also both the very disappointing performance of the unmodified HDPE and the HDPE modified with $CaCO_3$ particles of 3.5μm diameter with ligament thickness well above the threshold level of 0.8μm. A further important point is that the development of the toughening mechanism requires relatively modest levels of adhesion between particles and matrix to permit the required debonding for $CaCO_3$. In the case of the EOR rubbers which had a high level of adhesion the toughening mechanism could still operate as long as the particles cavitated internally, but when this did not occur below the respective T_g's of the rubbery particles, the toughening mechanism became inoperable.

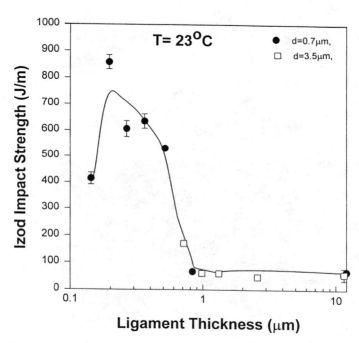

Figure 12   Toughness jump in HDPE/CaC0₃ blends at room temperature when average interparticle ligament thickness decreases below c.a. 0.8μm, in experiments with particle size 0.7μm where volume fraction of particles was varied to change the ligament thickness.

118

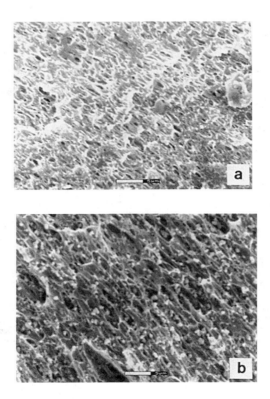

Figure 13 SEM micrographs of cryo-fracture surfaces along planes perpendicular to the fracture surface and parallel to the surface planes of Izod impact fracture specimens, showing the sausage shaped plastically elongated cavities in the carcass of PE blends subsequent to cavitation or decohesion of particles: a) blend containing EPDM particles; b) blend containing CaC0₃ particles. Areas shown are immediately under the fracture surface with the vertical directions being perpendicular to the fracture surface. Volume fraction 0.22 in each case.

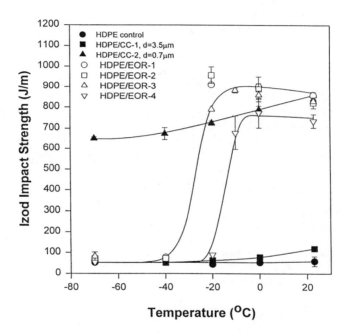

Figure 14 Temperature dependence of Izod impact energy in rubber particle containing PE blends showing embrittlement of the well adhered rubber particles which do not cavitate below their respective T_g's. In comparison $CaCO_3$ particle modified blends where particle debond toughness is temperature independent. Neat HDPE and HDPE blends with large particles exhibit only brittle behavior.

Finally, we summarize the results on HDPE in Figure 15. Homo-HDPE has a Young's modulus of 800 MPa at room temperature. Incorporation of rubbery particles with inter-particle ligament dimensions below the threshold, result in the familiar impressive toughening effect, but at a significant compromise in the modulus of the blend by about 60%. Incorporation of $CaCO_3$ particles at the same volume fraction and in a size range to give ligament thickness under the threshold value do similarly well in enhancing the Izod fracture energy of these blends, but with an added attractive improvement of the Young's modulus by about 50%. Increasing $CaCO_3$ particle sizes for the same volume fraction (increasing the ligament thickness) results in the same level of improvement of Young's modulus but in a drastic compromise on the toughness, since the critical ligament thickness requirement is exceeded.

Discussion

In the present communication we devoted our attention to the toughening of notch brittle semi-crystalline polymers like nylon and HDPE, where the relevant morphological size scale is in μms. For these, polymers in which large strain plasticity is almost exclusively a result of crystallographic shear processes, the toughness jump has been related to the attainment of an interparticle ligament dimension, based on the observation of Wu (22,23) for nylon blends. The same notion was found to be applicable to HDPE by Fu, et al. (36), and to iPP by Wu et al. (40).

Other toughening phenomena in semi-crystalline polymers are known to occur on the nanoscale where the inclusions are rubbery block copolymer domains (41,42). The mechanisms of this toughening process, also based on notions of percolation of material with enhanced molecular mobility in thin layers around the rubbery particles, is however, quite different and does not involve crystal plasticity.

Other toughening mechanisms are likely to be found as variants of the processes described in this communication. In all cases, however, the common denominator is a reliance on promotion of dissipative processes at reduced stress levels that delay or entirely avoid fracture processes emanating from extrinsic or intrinsic microstructural imperfections by the rather elementary principles outlined above.

Finally, we are aware of numerous other reports on toughening of glassy polymers with rubber particles where a threshold morphological scale is involved. A clear and well understood case is in the toughness of the PS/PB block copolymers where rubbery domains of spherical or rod shaped forms cavitate ubiquitously, resulting in very regular cellular PS carcasses of morphological wave length of the order of 20-40nm. In such quasi regular cellular structures at their particular scale, imperfections resulting from adventitious flaws are rendered inoperative, giving rise to a combination of high flow stresses and large strains to fracture with energy absorption in considerable excess of HIPS or ABS (13,14). Other amorphous systems often brought up as counter arguments to our specific mechanism are frequently ill

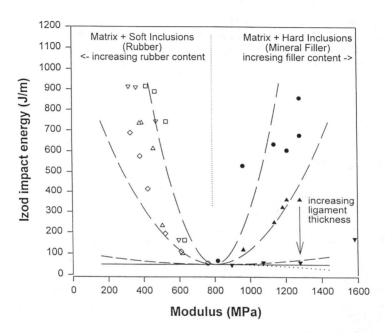

Figure 15 Summary of results of toughening of PE in blends utilizing either rubbery particles or CaCO₃ particles at 0.22 volume fraction. When the critical ligament thickness condition is satisfied in the blends both types of particles result in 10-15 fold increase in the fracture energy with the CaCO₃ particles and in addition resulting in a 20-50% increase in Young's modulus while the rubber blends compromise the Young's modulus by about 50%. With CaCO₃ particles when the particle size becomes too large and the interparticle ligament thickness exceeds the critical value, Izod toughness precipitously declines to background levels.

defined and lack supporting detail. They need to be carefully examined before convincing mechanisms can be stated.

Acknowledgment

This research has been supported by NSF/MRL and NSF/MRSEC through the Center of Materials Science and Engineering at MIT under Grants No. DMR-90-22933 and DMR-94-00334. Further support in the form of doctoral fellowships for OKM was provided by DuPont Co. for which we are grateful to Dr. D. Huang. Additional support was also received through an aid-to-education grant, again from the DuPont Co. For pedigreed materials and their characterization and for a multitude of other collaboration we are grateful to Dr. M. Weinberg also of the DuPont Co.

References

1. Kelly, A., Tyson, W.R., and Cottrell, A.H, *Phil. Mag.*, 1967, *15*, 567.

2. Rice, J.R., and Thomson, R., *Phil. Mag.*, 1974, *29*, 73.

3. McClintock, F.A., and Argon, A.S., "Mechanical Behavior of Materials" (Addison Wesley: Reading Mass), 1996.

4. Galeski, A., Bartczak, Z., Argon, A.S., and Cohen, R.E., *Macromolecules*, 1992, *25*, 5705.

5. Lin, L., and Argon, A.S., *Macromolecules*, 1992, *25*, 4011.

6. Argon, A.S., in "Advances in Fracture Research", edited by Salama, K., Ravi-Chandar, K., Taplin, D.M.R., and Rama Rao, P., (Pergamon: Oxford) 1989, vol 4, pp 2661-2681.

7. Argon, A.S, Cohen, R.E., and Muratoglu, O.K., in "Mechanics of Plastics and Plastic Composites-1995", edited by M.C. Boyce (ASME: New York), 1995, MD-Vol 68/AMD-Vol 215, pp. 177-196.

8. Argon, A.S., Cohen, R.E., Gebizlioglu, O.S., and Schwier, C.E., in "Advances in Polymer Science", edited by H.H. Kausch, (Springer: Berlin) 1983, vols 52/53, pp. 275-334.

9. Boyce, M.C., Argon, A.S., and Parks, D.M, *Polymer*, 1987, *28,* 1680.

10. Piorkowska, E., Argon, A.S., and Cohen, R.E., *Macromolecules*, 1990, *23*, 3838.

11. Dagli, G., Argon, A.S. and Cohen, R.E., *Polymer*, 1995, *36*, 2173.

12. Argon, A.S., Cohen, R.E., Jang, B.Z. and Vander-Sande, J.B., *J. Polymer Sci.*, 1981, *19*, 253.

13. Schwier, C.E., Argon, A.S. and Cohen, R.E., *Phil.Mag.*, 1985, *52*, 581.

14. Schwier, C.E., Argon, A.S., and Cohen, R.E., *Polymer,* 1985, *26,* 1985.
15. Gebizlioglu, O.S., Argon, A.S., and Cohen, R.E, *Polymer,* 1985, *26,* 579.
16. Gebizliglu, O.S., Argon, A.S., and Cohen, R.E., *Polymer,* 1985, *26,* 529.
17. Gebizlioglu, O.S., Beckham, H.W., Argon, A.S., Cohen, R.E., and Brown, H.R., *Macromolecules,* 1990, *23,* 3968.
18. Argon, A.S., Cohen, R.E., Gebizlioglu, O.S., Brown, H.R., and Kramer, E.J.,*Macromolecules,* 1990, *23,* 3975.
19. Argon, A.S., and Cohen, R.E., in "Advances in Polymer Science", edited by H.H. Kausch (Springer-Berlin) 1989, vols 91/92, pp. 301-351.
20. Argon., A.S., Cohen, R.E., and Mower, T.M, *Mater. Sci. Engng.,* 1994, *A176,* 79-90.
21. Kinloch, A.J., "Mechanics and Mechanisms of Fracture of Thermosetting Epoxy Polymers", in "Advances in Polymer Science", (Springer-Berlin) 1985, vol. 72, pp 43-67.
22. Wu, S., *Polymer,* 1985, *26,* 1855.
23. Wu, S., *J. Appl. Polym. Sci., 35,* 1988, 549.
24. Sue, H.-J., *J. Mater. Sci.,* 1992, *27,* 3098.
25. Borggreve, R.J.M., Gaymans, R.J., Schuijer, J., and Ingen Housz, J.F., *Polymer,* 1987, *28,* 1489.
26. Galeski, A., Argon, A.S., and Cohen, R.E., *Makromol. Chem.,* 1987, *188,* 1195.
27. Muratoglu, O.K., Argon, A.S., and Cohen, R.E., *Polymer,* 1995, *36,* 2143.
28. Muratoglu, O.K., Argon, A.S., Cohen, R.E., and Weinberg, M., *Polymer,* 1995, *36,* 921.
29. Parks, D.M., and Ahzi, S., *J. Mech. Phys.Solids.,* 1990, *38,* 701.
30. Lee, B.J., Argon, A.S., Parks, D.M., Ahzi, S. and Bartczak, Z., *Polymer,* 1993, *34,* 3555.
31. Muratoglu, O.K., Argon, A.S., Cohen, R.E., and Weinberg, M., *Polymer,* 1995, *36,* 4787.
32. Kumar, V., German, M.D., and Shih, C.F., "An Engineering Approach for Elastic-Plastic Fracture Analysis", (NP-1931, Research Project 1237-1 Report) Electric Power Research Institute: Palo Alto, CA., 1981.
33. Lin, L., and Argon, A.S., *Macromolecules,* 1994, *27,* 6903.
34. Rice, J.R., and Sorensen, E.P., *J. Mech. Phys. Solids,* 1987, *26,* 163.
35. Tzika, P., Boyce, M.C. and Parks, D.M., to be published.
36. Fu, Q., Wang, G., and Shen, J., *J. Appl. Polym. Sci.,* 1993, *49,* 673.
37. Bartczak, Z., Argon, A.S., and Cohen, R.E., *Macromolecules,* 1992, *25,* 5036.
38. Chacko, V.P., Karasz, F.E., Farris, R.J., and Thomas, E.L, *J. Polym. Sci-Phys*: 1982, *20,* 2177.
39. Muratoglu, O.K., Argon, A.S., Cohen, R.E., and Weinberg, M., *Polymer,* 1995, *36,* 4771.

40. Wu, X., Zhu, X., and Qi, Z., in "Deformation Yield and Fracture of Polymers" (The Plastics and Rubber Institute: London) pp. 78/1-78/4; *Ibid*, pp. 79/1-79/4, 1991.

41. Mukai, U., Gleason, K.K., Argon. A.S., and Cohen, R.E., *Macromolecules,* 1995, *28*, 4899.

42. Mukai, U., Argon, A.S., and Cohen, R.E., *Polym. Engng. Sci.,* 1996, *36*, 895.

43. Muratoglu, O.K., 1995, Ph.D Thesis in Materials Science and Engineering, Massachusetts Institute of Technology, Cambridge, MA.

Chapter 8

Deformation of Polypropylene–EPDM Blends

Reinoud J. Gaymans and Allard van der Wal

University of Twente, P.O. Box 217, 7500 AE Enschede, The Netherlands

Polypropylene-EPDM rubber blends with rubber content varying between 0 - 30 vol. %, were made on a twin screw extruder and their mechanical properties studied. These included the measurement of the notched Izod fracture energy and the notched tensile behavior as function of strain rate (0.003 - 300 s^{-1}) and also as a function of temperature (-40° - 140°C). During high speed testing, the deformation zone of the sample, warm up, and temperature development were followed using an infrared camera. The structure of the deformation zone was studied by scanning electron microscopy.

The blends were found to exhibit a sharp brittle-ductile transition and have high fracture energies in the ductile region. By increasing the rubber content, the brittle ductile transition shifted strongly towards the lower temperatures. The deformation process changed with temperature and test speed, the test speed effect possibly being due to a strong adiabatic heating in the fracture plane and therefore a thermal blunting mechanism is proposed.

Introduction

Polypropylene (PP) is a semi-ductile material which, when unnotched, has a high impact strength, and when notched, fractures in a brittle manner. With increasing temperature, the material becomes ductile at the brittle-ductile transition. For PP, the brittle-ductile transition temperatures (T_{bd}) on notched samples and measured in Izod are well above room temperature (1). The T_{bd} can be lowered by the incorporation of a rubbery dispersed phase; the rubber can be blended in by using an extrusion compounding process, or can be polymerized in. Although incorporation by polymerization is the usual industrial practice(2), the extrusion process is more flexible in studying the behavior of the blend as a function of morphology. PP is usually modified with EPR and EPDM rubber but can also be toughened by using SBS, SEBS, polybutadiene and polyisoprene(3-13).

With rubber modification, the modulus and the yield strength are reduced, the notched Izod energy is increased, and a considerable shift in the brittle-ductile transition temperature is observed. The function of the rubber is to cavitate and thereby change the stress state of the matrix in the vicinity of the cavity (*14-19*). An additional effect of cavitated particles concentration is a stress field overlap of neighboring cavitated particles(17,18,20,21). As a result of the cavitation, the plane strain condition ahead of a notch is gradually changed to a plane stress state and shear yielding of the matrix becomes more likely to occur. Shear yielding and crazing are competitive processes, the one with the lowest stress occurring. At the brittle-ductile transition, the yield stress crosses over with the craze stress. This transition state is sensitive to changes in material parameters and test conditions.

The notched Izod impact method is often used for studying the impact behavior of engineering plastics. With an increasing rubber concentration, the notched Izod fracture energies of PP increase (Figure 1)(*12*).

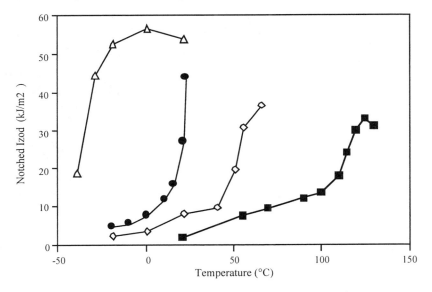

Figure 1. Notched Izod as function of temperature of PP/EPDM with varying rubber content (vol. %): ■, *0%;* ◇, *15%;* ●, *20%;* △, *30%. Reproduced with permission from reference 12.*

At low temperatures (-40°C), the samples fracture in a brittle manner and the fracture energy increases with increasing rubber content. At high temperatures, the samples deform ductile and are no longer fully broken. In addition, the T_{bd} transition is also shifted towards a lower temperature with an increasing rubber content. Samples that exhibit brittle fracture have: a relative low fracture energy (1-20 kJ/m2), stress whitening in the notch, a thin stress whitened layer adjacent to the fracture plane and a high fracture speed (> 200 m/s). Samples that fracture in a ductile manner have: a much higher fracture energy, a stress whitened zone adjacent to the fracture plane (1-2 mm thick) and a low fracture speed ($<$ 50 m/s). The matrix parameters have a direct effect on the impact properties of the blend. By increasing the molecular weight and decreasing the crystallinity of PP, the T_{bd} of the blend is decreased (11).

The morphology of the blend and in particular the rubber particle size, has an appreciable effect on the notched Izod behavior(22). Decreasing the rubber particle size shifts the T_{db} to a lower temperature, this effect is similar to that observed in the polyamide system. The function of the rubber particles is to cavitate; however, the cavitated particles should not become initiation sites for the fracture process. Therefore, the particles must be small enough such that they do not grow to a size where they are able to initiate a crack. It is expected that large particles form large cavities and thus are more likely to initiate fracture.

Ductile fracturing blends exhibit extensive plastic deformation. As most of the mechanical energy is dissipated as heat, a marked temperature increase takes place in the notch area(13). If the plastic deformation takes place under adiabatic conditions (at high strain rates), an even stronger temperature rise can be expected.

These ductile materials are generally studied using the notched Izod method. An additional, and very useful method is the single edged notch tensile (SENT) test (12). This instrumental method provides information on the fracture stress, strain development before crack initiation and strain development following crack initiation. In the notch, plastic deformation can take place prior to a crack being initiated. At the moment of crack initiation, stresses are likely to be at their maximum and at this point a considerable amount of elastic energy is stored in the sample. If this elastic energy is sufficient to lead to fracture of the sample, then the crack proceeds fast and the fracture type is brittle. If the elastic energy in the sample is insufficient to fracture the sample, more energy has to be supplied. The fracture speed is therefore much lower and the fracture proceeds in a ductile manner.

In this paper we have studied the influence of strain rate on the fracture behavior of PP-EPDM blends using the SENT test method.

Experimental

The materials used encompass a polypropylene homopolymer (Vestolen P7000, Vestolen GmbH) with an MFI of 2.4 (230 °C, 21.6 N) and an EPDM rubber (Keltan 820, DSM). A series of blends of varying rubber content were prepared by diluting a PP-EPDM (60/40 vol. %) extrusion master blend and rectangular bars and dumbbell-shaped specimens prepared by injection molding. The blending and injection molding processes have previously been described in detail (12).

The SENT tests were carried out on notched bars using a Schenck servo-hydraulic tensile machine. The injection-molded bars (ISO 180/1, 74x10x4 mm) had a milled, single-edge, 45 ° V-shaped notch (depth 2 mm, tip radius 0.25 mm). The clamp length of the samples was 45 mm. The Schenck tensile apparatus is specifically designed for high speed testing, in that the studied clamp speeds range from 10^{-5} to 10 m/s. At high test speeds, a pick-up unit was used, thus allowing the piston to reach the desired test speed before specimen loading. To minimize inertia effects, all moving parts were made of titanium. Force and piston displacement were recorded using a transient recorder (sample rate 2 MHz) and following test completion, all results were downloaded to a computer. All measurement were carried out in five fold.

The temperature rise during fracture was measured by an infrared camera (IQ 812 system from Flir Systems) (*13*), the signal from which was sent to a computer.

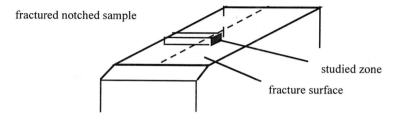

fractured notched sample

studied zone

fracture surface

Figure 2. Sample position

The mechanism of deformation of the fracture zone was studied by post-mortem using scanning electron microscopy (SEM), the sample position studied is shown in Figure 2.

The area of interest is perpendicular to the fracture surface and runs parallel to the crack growth direction. Samples were taken from the specimens using a fresh razor blade and then smoothed down on a polishing machine to avoid deformation during the trimming procedure. The finishing preparation was carried out by sectioning at -100 °C with a diamond knife and the final surface sputter-coated with a gold layer before analysis and studied using a Hitachi S-800 field emission SEM.

Results and Discussion

The studied PP-EPDM blends were prepared by extrusion blending, the particle sizes obtained are shown in Table 1.

Table 1. Particle size in PP-EPDM extrusion blends

Rubber Content	5%	10%	20%	30%	40%
Dw (μm)	0.50	0.67	0.69	0.86	0.92

The particle size increases with increasing rubber content, however such a particle size difference is known to have only a small effect on the brittle-ductile transition(*22*). The tensile and physical properties of the blends used have previously been reported(*12*).

Fracture behavior as a function of test speed was studied by using a single-edge notched tensile (SENT) test. The influence of test speed was examined by varying the clamp speed from 10^{-5} to 10 m/s. With these test speeds, a clamp length of 45 mm and a notch, the observed apparent strain rates in the elastic loading part are in the range 0.000275 - 275 s^{-1}.

The fracture process can be divided into two stages, these being 'initiation' and 'crack propagation'. During 'initiation', the stress increases at the site of the notch tip, but is still too low to allow crack propagation. Crack propagation begins at or past the stress maximum in the stress displacement curve. We found that PP deformed in this SENT test at room temperature, remained brittle over the whole strain rate range.

The fracture energy at 27.5 s^{-1} (1m/s) was studied as function of test temperature at different EPDM contents. When using the SENT test, the samples were always fully broken, the fracture energies being reliable up to high temperatures. S-type curves were obtained and with increasing rubber content these S-type curves are shifted to lower temperatures (Figure 3).

At low temperatures the samples fractured in a brittle manner having a low fracture energy. At high temperatures they fractured in a ductile manner following a sharp brittle-ductile transition. The fracture energies in SENT were considerably higher than those obtained in Izod. In tensile loading (SENT), more material was plastically deformed and the stress-whitened zone was thicker, particularly at the end of the fracture path. With an increasing rubber content, the S-curves were shifted to lower temperatures. In the brittle region (-30°C), fracture energies increased with increasing rubber content (Figure 4).

At -30°C the fracture is in all these cases brittle and this increase in fracture energy is due solely to the increased deformation in the notch. As with increasing rubber content deformation in the notch area becomes stronger and the fracture energy increases. In this brittle region, the fracture energies as measured in SENT are comparable to the fracture energies as measured in notched Izod. At high temperatures (120°C) the fracture is in all cases ductile and the fracture energies decreased with increasing rubber content. The function of rubber is to cavitate and change thereby the stress state. The high fracture energy comes from the plastic deformation of the matix. This plastic deformation is ennabled by the rubber particles. Ones a material fractures ductile at a particular temperature than more rubber does not futher enhances the fracture energy it even decreases the fracture energy as the matrix content is decreased with increasing rubber content. Surprisingly, the 1% and 5 % blends have higher fracture energies than neat PP.

The ductile region (past the T_{bd}) the fracture energy decreases with increasing temperature (Figure 3) is due to the corresponding strong decrease in yield strength with temperature.

Ones a material fractures in a ductile manner a futher lowering of the yield strength doesnot so much enhance the fracture strain but lowers the fracture energy. The brittle-ductile transition temperature was found to decrease with increasing rubber content (Figure 5). This decrease is similar to that observed in the notched Izod test; therefore the SENT test at 1 m/s is comparable to the notched Izod. T_{bd} decreased steadily with increasing rubber content. Therefore it appears that a small percentage of rubber is insufficient to relieve all of the volume strain. The glass transition of PP (at 5°C) had little effect on this T_{bd} shift. Below its T_g, PP has a more brittle behavior but this is apparently balanced by a more effective toughening by the rubber. Below its T_g, the PP has a lower Poisson-ratio which makes the material more susceptible to crazing, however, if the matrix has a low Poisson-ratio the rubber cavitation takes also place more easily.

The SENT studies at different strain rates were carried out in two parts. Firstly, a fixed rubber concentration was used (10 % EPDM) and the temperature varied, secondly, the rubber concentration was varied whilst the temperature was held constant at room temperature.

The stress displacement curves as obtained by SENT on a 10 vol. % blend at low and high strain rates are shown (Figure 6).

At a strain rate of 27.5 s^{-1}, the blends fractured in a brittle manner at 0° and 40°C

130

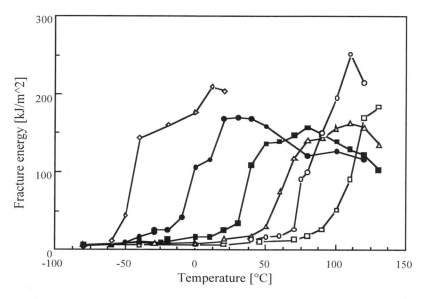

Figure 3. SENT fracture energies as function of temperature measured at 27.5 s⁻¹ (1m/s) of PP/EPDM with varying rubber content (Vol. %): ☐,0 ; ◯, 5; △, 10; ■, 20, ●, 30; ◇, 40. *Reproduced with permission from reference 12.*

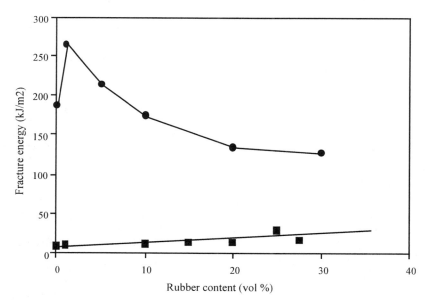

Figure 4. SENT fracture energies as function of rubber content measured at 27.5 s⁻¹ (1m/s) and at different temperatures: ■, - 30°C; ●, 120°C.

and in a ductile manner above 60°C. The fracture stress decreased strongly with increasing temperature. At temperatures of 60°C and higher, considerable yielding occurred in the notch and during fracture. At a strain rate of 0.0275 s^{-1}, the samples became ductile at 20°C and behaved in a fully ductile manner at 50°C. In this 20°-50°C temperature range, the fracture energy in the propagation region is due to deformation with shear lips. The T_{bd} for the 0.0275 s^{-1} strain rates was measured as a function of the rubber content (Figure 5). At this strain rate, the T_{bd} decreased with increasing rubber content, such that the T_{bd} curve as a function of the rubber content was 30°C lower than that in the 27.5 s^{-1} SENT results. The more brittle behavior at higher strain rates is as expected.

The stress displacement curves obtained by the SENT test for the 30 Vol. % blend at varying test speeds are shown in Figure 7.

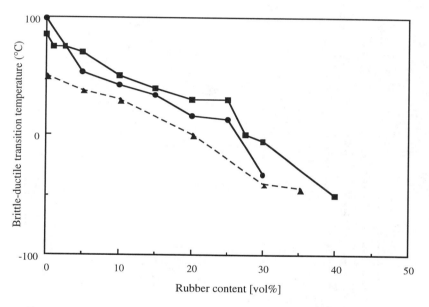

Figure 5. The brittle-ductile transition temperature as function of rubber content: ●, *Izod;* ■, *SENT 27.5 s^{-1} (1m/s);* ▲, *SENT 0.0275 s^{-1} (1 mm/s). Reproduced with permission from reference 12.*

The maximum stress of this 30% blend increased with increasing strain rate and considerable plastic deformation was found to take place during crack propagation. Throughout the entire speed range used in the test, this 30% blend fractured in a ductile manner.

For polypropylene and the 30% blend, the maximum stress as function of test speed is shown in Figure 8.

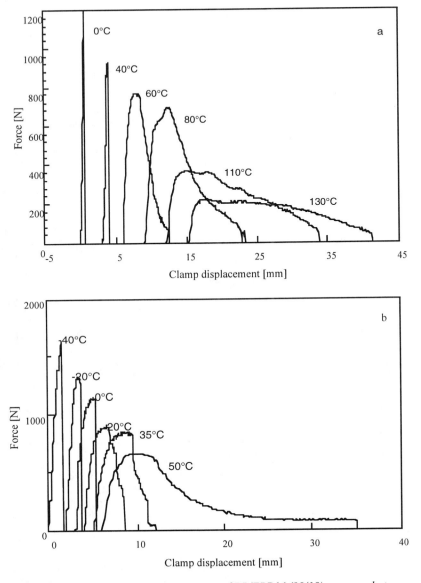

Figure 6. SENT stress displacement curves of PP/EPDM (90/10) measured at different temperatures: a, 27.5 s⁻¹; b, 0.0275 s⁻¹.

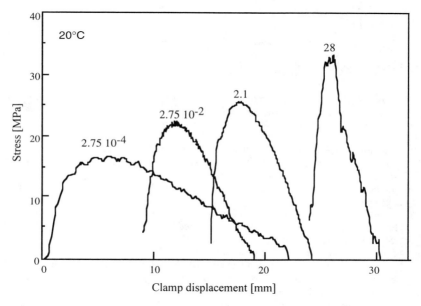

Figure 7. SENT stress displacement curves of PP/EPDM (70/30) measured at different strain rates. Reproduced with permission from reference 13.

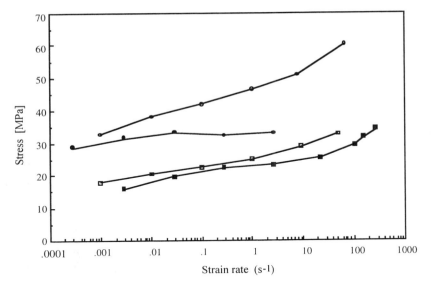

Figure 8. The yield stress in tensile and the maximum stress in SENT as function of strain rate: ○, PP tensile; □, PP SENT; ●, PP/EPDM (70/30) tensile; ■, PP/EPDM (70/30) SENT. Reproduced with permission from reference 13.

For comparison the measured yield stresses for unnotched specimens are also given. At a low strain rate (10^{-3} s^{-1}), PP has a maximum stress that almost equals the yield stress. This suggests that the whole cross section ahead of the notch sustained a yield stress and therefore that the notch must have been blunted. At intermediate test speeds, the maximum stress of the PP curve dropped to a level well below the yield stress, therefore crack propagation started before the entire cross section was bearing a yield stress. If the material fractured before any yielding in the notch had taken place, then it would be expected that there would be a strong decrease in the fracture stress. This was not observed in PP at 20°C or even -40°C

at the strain rates tested. Over almost the whole strain rate region, the 30% blend has a stress that equals the yield stress. At very high strain rates, a strong increase in stress is observed, a similar increase can be seen in the unnotched tensile results. This increase seems to be typical for materials that exhibit a strong plastic deformation in the notch.

The initiation energy (IE) was found to be high at low test speeds and subsequently decrease with increasing test speed (Figure 9a). At very high test speeds, the IE appears to stabilize or even increase. The blends tested exhibited a strong plastic deformation before the initiation of a crack over the entire speed range used. The amount of plastic deformation in the notch before crack initiation was found to decrease with a increased strain rate, however at the higher strain rates, stabilization seemed to occur. This stabilization in the deformation in the notch suggests an enhanced notch blunting effect. Therefore, the amount of plastic deformation in the notch has an effect on the stress concentration in the notch.

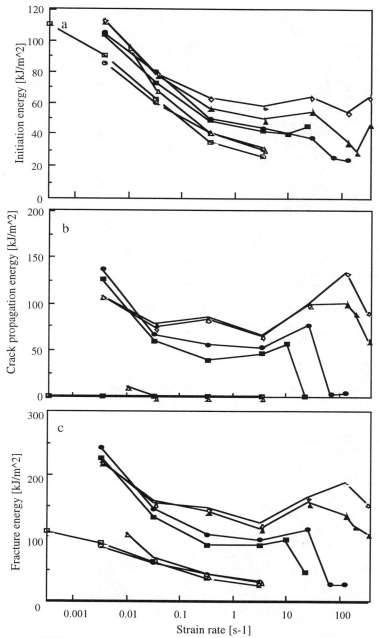

Figure 9. SENT fracture energies as function of strain rate at different EPDM contents (vol%): ■*, 0;* ○*, 1;* △*,5;* ■*,,15;* ●*, 20;* ▲*, 30;* ◇*, 40. Reproduced with permission from reference 13.*

136

The crack propagation energy (CPE) (Figure 9b) curves are complex. At low strain rates, the CPE decreased with increasing strain rate. This was as expected as by increasing the strain rate, the fracture strain decreases. Surprisingly, at intermediate strain rates, the CPE is increased with increasing strain rate. At very high strain rates, it decreases again. The results suggest that at low strain rates with increasing test speeds, the materials turn brittle. However, in all samples, at intermediate strain rates (0.2 - 2 s^{-1}) this trend is reversed, such that the materials become less brittle. At even higher strain rates, the materials again turned brittle. It seems as if the at high test speeds the fracture process proceeds diffrently from at low speeds and therby shifting the trend of decreasing fracture energies with increasing strain rates to higher strain rates. This change at high strain rates is probably the result of an adiabatic plastic deformation accompanied by a strong warming of the sample. A similarly complex behavior of the CPE with strain rate was found in rubber modified nylon, this has been be explained as a thermal blunting effect (*23*). The rubber content had a strong influence on when the blends started to behave in a brittle manner. The higher the rubber content, the higher the strain rate at which the blends turned brittle. The turning ductile at 0.2 - 2 s^{-1} suggests that at this strain rate, the deformation process undergoes a change.

The total fracture energy (Figure 9c) also exhibits a complex behavior as a function of strain rate. The fracture energy is almost equally dependent on the initiation and the propagation parts of the deformation process. Over the whole strain rate range at this temperature, the fracture energy strongly increased with an increasing rubber concentration. This is due to the fact that neat PP fractures in a brittle manner over the whole strain rate range whilst the 30 % blend is fully ductile over the same range. The other blends exhibit behaviors in-between those of the neat PP and the 30 % blend.

In the blends, plastic deformation clearly takes place in the notch and the process proceeds by ductile fracture during crack propagation. With this strong plastic deformation, a considerable temperature rise can be expected at high test rates. The temperature increase during SENT was studied by using an infrared camera (*22*). The spatial resolution of the camera is 130 μm whilst the strongest deformation can be observed in a 50 μm thick layer adjacent to the fracture surface. A typical pattern is shown in Figure 10, the highest temperatures being recorded just ahead of the notch. The maximum wall temperature increased with an increasing strain rate to 90°C at 10m/s. The poor spatial resolution of the infrared camera does not allow the study of the very thin (50μm) layer adjacent to the fracture plane, this is the region where the strongest plastic deformation is expected.

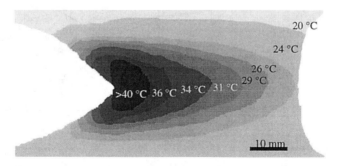

Figure 10. Temperature profile of PP/EPDM (70/30) sample deformed at 0.027 s^{-1}. Reproduced with permission from reference 13.

To study the asymmetric deformation process in fractured notched samples, SEM micrographs were made of the fracture zone next to the fracture surface in the middle of the samples. We examined the effect of strain rate on the structure of the fracture zone in a 30 vol. % blend that fractured in a ductile manner (Figure 11).

At low strain rates (0.0275 s^{-1}), on a fracture sample taken from an area adjacent to the fracture surface, strongly elongated cavities could be seen. For cavities at the fracture surface the length/width ratio was 15, thus suggesting a draw ratio for PP of 15. On increasing the samples distance from the fracture surface, the cavities become more rounded. At higher strain rates ($\geq 0.2 \text{ s}^{-1}$), the deformation of the cavities appears to be less than that seen at low strain rates, in that there are no longer any cavities present in the layer adjacent to the fracture surface. In the region just ahead of a crack, a layer without cavities can also be seen. This suggests that in PP-EPDM blends, the elongated cavities disappear at this high strain rate. The cavities seem to have relaxed in the region next to the fracture plane. This may be due to the temperature increase at these high strain rates. Up to approximately 0.275 s^{-1}, the cigar-shaped voids extend to the fracture surface, the thickness of the strongly deformed layer being about 25 μm. The appearance of the relaxation layer at high test speeds is also observed in nylon 6-rubber blends(23,24). In these blends, the thickness of the relaxation layer was found to be 3-5 μm. The appearance of a relaxation layer in polypropylene-EPDM blends at high strain rates and coincides with the CPE increase with the strain rate (Figure 9c).

Conclusions

By using the instrumented SENT method, considerable insight into the deformation behavior of PP-rubber blends can be obtained. Notch deformation of PP-EPDM blends, as a function of strain rate is complex. At low strain rates, the crack propagation energy (CPE) was found to rapidly decrease with increasing strain rate. However, at intermediate strain rates ($> 0.2 \text{ s}^{-1}$), the CPE increased with increasing test speed. At these strain rates, the structure of the deformation zone (of the 30% blend) adjacent to the fracture surface changes and the samples exhibit a layer that must have been relaxed. This suggests that during plastic deformation at high strain rates a highly elastic material is present ahead of the notch and that this layer may blunt the notch. This is termed a thermal blunting process.

Thus two observed processes occur. At low strain rates, a fracture process accompanied by strong plastic deformation occurs both during initiation and as propagation proceeds. At high strain rates, a plastic deformation process occurs where there is formation of a elastic layer ahead of the notch front. This thermal blunting process that is taking place at high strain rates considerably enhances the toughness of the blend, if this process did not occur, these blends could well be brittle at the strain rates used.

138

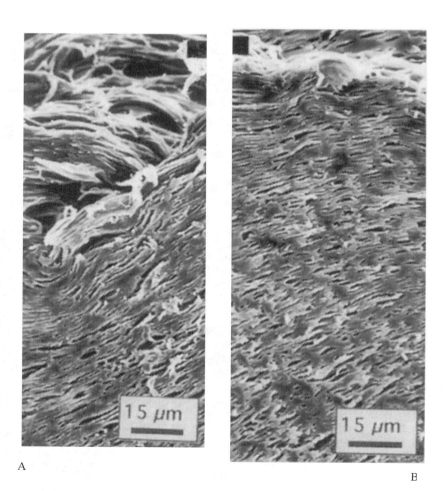

Figure 11. Structures of the fracture zone at different strainrates: A, 0.0275 s⁻¹; B, 0.275 s⁻¹; C, 2.75 s⁻¹; D, 22 s⁻¹. Reproduced with permission from reference 12.

C

D

Figure 11. *Continued.*

References

1. Van der Wal, A; Mulder, J.J.; Thijs, H.A.; Gaymans R.J., *Polymer* **1998**, *39*, 5467.
2. Gali, P.; Haylock, J.C.,*Macromol. Chem. Macromol Symp.*, **1992**, *63*, 19.
3. Ramsteiner, F. *Polymer* , **1979**, *20*, 839.
4. Karger-Kocsis, J., *Polypropylene structure and properties*, Chapman Hall, London, Vol 2, 1995.
5. Chou, C.J.; K. Vijayan, K.; Kirby, D.; Hiltner, A.; and Baer, E., J. *Mat. Sci.* **1988**, *23*, 2521.
6. Greco, R.; Mancarella, C.; Martuscelli, E.; Ragosta, G.; Jinghua, J., *Polymer* **1987**, *28*, 1929.
7. Jang, B.Z.; Uhlmann, R.D.; Vander Sande, J.B., J. *Appl. Polym. Sci.* **1985**, *30*, 2485.
8. Michler, G.H., Kuststof-Mechanik: Morphologie, Deformations- und Bruchmechanismen, Carl Hanser Verlag, Munich, 1992.
9. Ramsteiner, F. *Acta Polymerica*, **1991**, *42*, 584.
10. Inoue, T.; Suzuki, T., J. *Appl. Polym. Sci.* **1996**, *59*, 1443.
11. Van der Wal, A.; Mulder, J.J.; Oderkerk, J.; Gaymans, R.J., *Polymer*, **1998**, *39*, 6781.
12. Van der Wal, A.; Nijhof, R.; Gaymans, R.J., *Polymer*, **1999**, *40*, 6031.
13. Van der Wal, A.; Gaymans, R.J., *Polymer*, **1999**, *40*, 6045.
14. Bucknall, C.B., *Adv. Polym. Sci.*, **1978**, *27*, 121.
15. Donald, A.M.; Kramer, E.J., J. *Mat. Sci.*, **1982**, *17*, 1765.
16. Ramsteiner, F.; Heckmann, W., *Polym. Commun.* **1985**, *26*, 199.
17. Yee A.F.; Pearson, R.A., J. *Mat. Sci.*, **1986**, *21*, 2462.
18. Borggreve, R.J.M.; Gaymans, R.J.; Schuijer, J., *Polymer*, **1989**, *30*, 71.
19. Dijkstra, K.; van der Wal, A.; Gaymans, R.J., J. *Mat. Sci.*, **1994**, *29*, 3489.
20. Dijkstra, K.; ten Bolsher, G.H., J. *Mat. Sci.*, **1994**, *29*, 4286.
21. Lazzeri, A.; Bucknall, C.B., *Polymer*, **1995**, *36*, 2895.
22. Van der Wal, A.; Verheul, A.J.J.; Gaymans, R.J., *Polymer*, **1999**, *40*, 6057.
23. Dijkstra, K.; Gaymans, R.J., J. *Mat. Sci.*, **1994**, *29*, 323.
24. Janik, H.; Gaymans, R.J.; Dijkstra, K., *Polymer*, **1995**, *36*, 4203.

Chapter 9

The Role of Impact Modifier Particle Size and Adhesion on the Toughness of PET

Thomas J. Pecorini and David Calvert

Eastman Chemical Company, P.O. Box 1972, Kingsport, TN 37662

PET was made supertough by melt blending with 20 weight percent of a reactive terpolymer (ethylene - methyl acrylate – glycidyl methacrylate). Supertoughness was obtained using both amorphous and semicrystalline PET matrices. Two distinct toughening mechanisms were identified: massive shear yielding in the matrix when the dispersed particle size was less than 200 nanometers, and multiple crazing in the matrix when the particle size was larger. Samples that contained only 10% of the reactive terpolymer, or a non-reactive copolymer, were brittle. Interparticle distance alone was insufficient to explain these results. Instead, the improved toughness could be explained using a cavitation model.

Two of the most important variables known to affect the toughness of impact modified engineering thermoplastics are the particle size and amount of the impact modifier. In general, higher quantities of impact modifier (15-20%) and smaller particle size (0.2 – 0.5 microns) are required to induce "supertoughness" in pseudo-ductile polymers such as nylon, polycarbonate and PBT (1). Several studies have combined these two variables into a single parameter known as the "interparticle distance" (ID) and have attempted to correlate this parameter with impact behavior (1-10). The interparticle distance can be calculated as

$$ID = [(\pi/6\phi)1/3-1]dw$$

where ϕ is the volume fraction of impact modifier and dw is the weight average particle size. Theoretically, there is sufficient connectivity between ligaments to allow continuous shear yielding in the matrix during macroscopic deformation in blends with interparticle distances below a critical value. This continuous shear yielding leads to supertoughness.

While this theory does successfully explain much data, there are several problems inherent in many of the studies. For example, the role of particle cavitation is not rigorously addressed, although it is often assumed that cavitation is a precursor to shear yielding. Also, many systems have been investigated using only a constant volume fraction of impact modifier whereby any correlation with interparticle distance is merely a correlation with particle size. In addition, the effect of particle size is often confounded with the effect of particle-matrix adhesion since good adhesion during compounding is usually required to produce the small particle sizes needed to obtain good toughness. Thus, few studies have been able to compare the role of impact modifier adhesion on toughness using particles of similar size.

More significantly, one research group has recently proposed that the effect of interparticle distance on toughening only occurs when the matrix polymer is semicrystalline (11,12). This group attributes the interparticle distance effect to a preferential alignment of crystalline lamellae into a favorable shear direction. Still another research group suggests that the observed correlations of toughness with interparticle spacing are merely coincidental. They claim that the true effect of volume fraction and particle size is to alter the interacting stress conditions that lead to cavitation of the impact modifier, shear yielding of the matrix and crazing of the matrix (13-15).

This paper attempts to address several of these issues by examining an impact modified polyethylene terephthalate (PET). Both non-functional and functional impact modifiers can be blended into PET in order to vary the amount of particle-matrix adhesion. It is then possible to vary the particle size of the functional impact modifier by adding a catalyst to the blend. This technique can produce a range of particle sizes, including intentionally large particles with good adhesion. Furthermore, PET is an interesting material to study as it can be molded amorphous, but also subsequently crystallized. Therefore, both amorphous and crystalline matrix states can be examined using identical impact modifier morphologies.

One functional group that reacts rapidly with polyester during compounding is epoxy (16,17). Several authors have reported using copolymers of EPDM with glycidyl methacrylate (GMA), ethylene with GMA (E/GMA) or terpolymers of ethylene, an acrylate, and GMA to toughen polyesters (18-24). Typically it is reported that these reactive impact modifiers toughen polyester more effectively than do their non-reactive equivalents. Nonetheless, a direct comparison of toughness to particle size is not found in these reports. Several authors also have noted that the ability of these reactive polymers to toughen polyesters can be enhanced through the use of catalysts (23-25). Presumably, these catalysts affect toughness by altering the particle size of the impact modifier during compounding, although there may be some effect associated with crosslinking the rubber. Again, no direct correlation of toughness with particle size was reported. In this present study, we will use either an ethylene-methyl acrylate (E/MA) copolymer or an ethylene-methyl acrylate-glycidyl methacrylate (E/MA/GMA) terpolymer, compounded in the presence of several catalysts, to study the effect of particle size and adhesion on the toughness of both amorphous and crystalline PET.

Experimental

The impact modifiers used in this work were a random copolymer of ethylene and 24 weight percent methyl acrylate (non-reactive), and a random terpolymer of ethylene, 24% methyl acrylate and 8% glycidyl methacrylate (Reactive). Either 10 or 20 weight percent of these impact modifiers were blended into a standard PET homopolymer. Three different catalyst systems (A,B and C) were used during preparation of the blends. The weight average molecular weight of the PET in all compounded samples, measured on the molded bars, was 39,000. Crystallization of the samples did not change their molecular weight.

Samples were compounded on a Werner Pflighter 30mm twin screw extruder at 250 RPM, a 285°C melt temperature, and 25 pounds/hr using a general purpose screw with some additional mixing elements. The compounded samples were molded on a Toyo Plastar injection molding machine with a 280°C barrel temperature. The mold was chilled using antifreeze to −5°C. The samples emerged from the mold amorphous, even in the center of the bar, as confirmed by both DSC and TEM analyses. Crystalline bars were generated by annealing the molded bars in an oven at 120°C for 3 hours.

Imaging and sizing of impact modifier particles was performed using Scanning Transmission Electron Microscopy (STEM). Samples were sectioned at a thickness of 100nm on a cyro-ultramicrotome operated at −105oC and observed on a Philips CM12 TEM/STEM operated at 80kV accelerated voltage. Particle Size measurements were made by digital image analysis of the STEM images. Three fields of 1000x800 pixels were analyzed for each sample. Field width was 17.2 x 13.8 microns; pixel resolution: 58.1 pixels/micron. Standard field flattening and segmentation algorithms were applied and particle sizes were reported out as circular area equivalent diameters.

Fractography was performed using Scanning Electron Microscopy (SEM). Specimens were prepared from Charpy impact bars that contained two notches, equally spaced away from the loading line of the striker. Both notches initiated damage, but only one crack propagated to failure. The non-fractured crack region was polished in cross section on a cyro-ultramicrotome operated at −105oC. Alternatively, a cross section of the damage region was prepared by freeze splitting a previously tested Charpy bar half. These specimens were coated with a 30nm layer of gold palladium and observed on a LEO Gemini FEG- SEM operated at a 3kV accelerating voltage.

Izod impact properties were measured according to ASTM standard D256 using test method A. Standard 3.2 mm thick samples containing a 0.25 mm radius notch were tested at both 23°C and 0°C. Impact data was also obtained according to test method B (3 point bend Charpy) at 23°C using sharply notched 6.4 mm thick samples. This latter condition was run in order to increase the degree of constraint upon the samples (i.e. more plane strain at the notch tip). Tensile properties were measured according to ASTM standard D638 using standard Type I molded tensile bars tested at 23°C and 50 mm/min.

Results and Discussion

As anticipated, the morphology of each of the blends was strongly affected by the reactivity of the impact modifier and the catalyst used. Figures 1 and 2 shows micrographs for the blends containing 20% impact modifier. Figure 3 shows the particle size distributions of these blends, plotted on a log-normal particle size scale. Note that the morphology of the blends containing the non-reactive impact modifier was not affected by catalyst type, whereby only one representative micrograph is presented. It is apparent from these micrographs that the reactive impact modifier, when blended in the presence of catalysts A and B, produced much smaller particles than the non-reactive material. These morphologies are similar to what is observed when reactive impact modifiers are compared with non-reactive impact modifiers in nylon systems. The chemical reaction between the impact modifier and matrix dramatically lowers the interfacial tension between the two normally immiscible phases, allowing the impact modifier to develop into a fine, stable dispersion through the shear forces present in compounding. However, when blended in the presence of catalyst C, the reactive impact modifier formed into much larger particles, comparable in size with the non-reactive system. It is believed that, in addition to allowing reaction with the PET, catalyst C may have promoted crosslinking within the reactive particle, preventing good dispersion.

For comparison with mechanical properties, the weight average particle sizes for the blends containing 10% and 20% impact modifier are listed in Table 1, along with interparticle distances calculated from the equation provided in the introduction. Note that this equation assumes that all particles are spherical and, therefore, is not strictly applicable to the non-spherical particles found in the Reactive-C blend. Nonetheless, the interparticle distance calculation for this blend is accurate enough for a qualitative comparison with the other systems. Note that the appearance of the impact modifiers in the 10% blends was not much different from that found in the 20% blends although the average particle sizes for the 10% blends were slightly smaller. Particle size and appearance was not found to change upon crystallization of the matrix.

Tensile properties, listed in Table 1, show that all the reactive impact modifier blends exhibited improved elongation to break values compared the non-reactive blend. Higher loadings further improved the break elongation. Furthermore, the impact modified blends all possessed yield strength values below that of the neat materials in accordance with the equation $\sigma_{yb} = (1-x\phi)\sigma_m$ where σ_{yb} and σ_m represent the yield strengths of the blend and matrix respectively; x is a constant. In the reactive systems, x=1.375, a value observed in other systems (13). The matrix yield strengths of amorphous and crystallized PET are 50 MPa and 75 MPa respectively.

Impact data is also listed in Table 1, measured on samples with both amorphous and crystalline matrices. NB represents a no-break notched Izod result, with energy in excess of 1000 J/m. Only one set of data is provided for the non-reactive blends as the catalyst type was found to not affect the properties of this system. For comparison, the 23C notched Izod impact value for neat PET is only 45 J/m. In

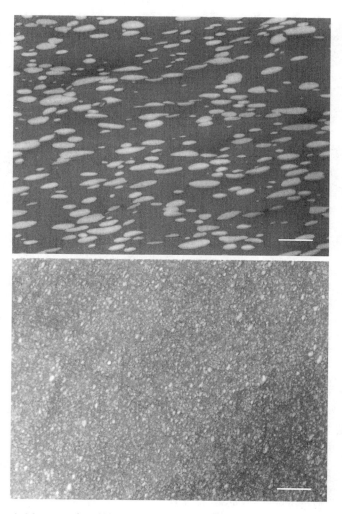

Figure 1. Micrographs of blends containing 20% impact modifier. The bar length equals 2 microns. Top – non-reactive. Bottom – Reactive-A.

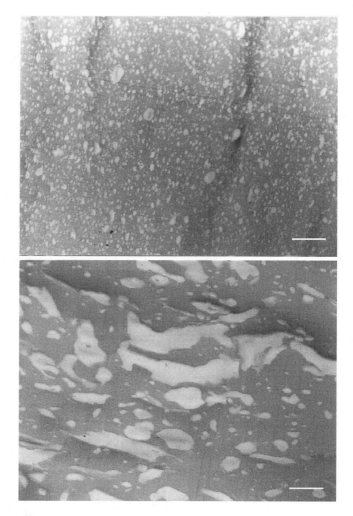

Figure 2. Micrographs of blends containing 20% impact modifier. The bar length equals 2 microns. Top – Reactive-B. Bottom – Reactive-C.

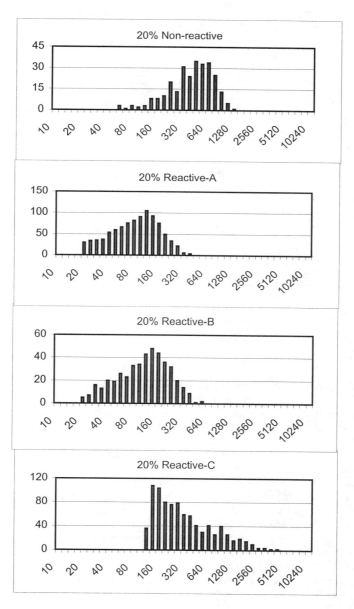

Figure 3. Particle size distributions for the 20% blends (nanometers).

general, only the reactive impact modifier was successful in improving the toughness of PET, as noted in prior studies. These improvements were only observed with 20% loadings. Toughness improvements were observed with both the amorphous and crystalline matrices, and in both the Izod and Charpy bars.

Table 1 – Particle Size and Mechanical Properties

	Non-Reactive	Reactive – A	Reactive – B	Reactive – C
Weight %	10% / 20%	10% / 20%	10% / 20%	10% / 20%
Particle size (nm)	454 / 530	160 / 130	180 / 170	560 / 970
ID (nm)	256 / 140	90 / 33	100 / 45	320 / 250
Amorphous				
Yield Strength (MPa)	47 / 38	42 / 33	43 / 32	41 / 29
Break Elongation (%)	44 / 52	180 / 330	190 / 360	427 / 363
23°C N Izod (J/m)	50 / 44	94 / NB	63 / NB	123 / NB
0°C N Izod (J/m)	43 / 42	71 / NB	63 / NB	89 / NB
23°C Charpy (J/m)	36 / 36	61 / 88	41 / 915	71 / 517
Crystallized				
Yield Strength (MPa)	67 / 56	60 / 46	60 / 47	58 / 40
Break Elongation (%)	8 / 7	16 / 65	9 / 46	18 / 68
23°C N Izod (J/m)	35 / 45	118 / 659	71 / NB	112 / 355
0°C N Izod (J/m)	36 / 44	85 / 351	72 / 814	86 / 292
23°C Charpy (J/m)	34 / 34	62 / 562	41 / 797	71 / 128

Interparticle Distance

Notched Izod values of 3.2 mm samples with a standard notch for all samples shown in table 1 are plotted as a function of interparticle distance in Figure 4. (No-break results are plotted with values of 1000 J/m). The Charpy data of 6.4 mm samples with a sharp notch is shown in Figure 5. (Note that all of the Charpy bars tested failed as complete breaks.) Several interesting trends are observed in these plots.

First of all, it was possible to attain supertoughness in these blends using both amorphous and crystallized matrices. The notched Izod values for the amorphous samples were often higher than their crystalline counterparts at both 23C and 0°C, but both sets of samples provided very similar trends of notched Izod with interparticle distance. This trend is still apparent even when tested under the more severe plane strain condition found in the sharply notched 6.4 mm thick charpy bars, although the toughness of the amorphous Reactive-A sample decreases significantly with the increased constraint. Nonetheless, the effect of interparticle distance on toughness is clearly not unique to semicrystalline matrices.

In general, the curve fit shown for the majority of the data in Figure 4 is typical of the relation between interparticle distance and notched Izod toughness as reported

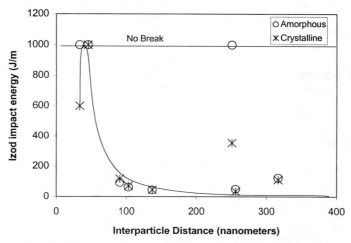

Figure 4. Notched Izod data for the samples as a function of interparticle distance

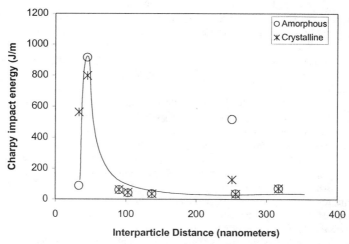

Figure 5. Charpy data for the samples as a function of interparticle distance

in the literature. Toughness increases as the interparticle distance decreases to a critical size, but becomes lower again as the distance becomes too small. Based on this curve, the critical interparticle distance for PET blends would be 50 nanometers. However, the samples made using 20% reactive impact modifier with catalyst C show good toughness even though their interparticle distance is 250 nanometers. Indeed, the blend containing 20% Reactive-C was tougher than the blends containing 10% Reactive-A and Reactive-B, even though the blend with 20% Reactive-C has a larger interparticle distance. In contrast, the non-reactive impact modifier (also with a large interparticle distance) was not effective in toughening PET at either 10% or 20% loading. These data strongly suggest that adhesion is more important than particle size in imparting good toughness to these blends, and that there is a distinct separation of amount and size in this system that can not be explained solely using interparticle distance theory.

Fractography

The role of particle size and adhesion is made clearer through fractographic examination to reveal the failure mechanisms that operate in impact modified PET. Figures 6, 7 and 8 show cross sections near the crack tip in Charpy bars of the amorphous 20% Reactive-B, Reactive-C and non-reactive blends. Figure 6-Top shows the massive matrix deformation and impact modifier elongation in the Reactive-B blend that are typical of supertough behavior. This image supports the theory that the role of small particles and a small interparticle distance is to induce high toughness by promoting massive shear yielding in the matrix.

In contrast, Figure 6-Bottom shows very little matrix deformation around the much larger, poorly bonded particles in the non-reactive blend. In this image, the crack is propagating from the razor notch at the lower right up to the upper left of the image. What little deformation exists occurs only as local expansion of voids around debonded particles. A cross sectional freeze fracture of the crack tip region (Figure 7-Top) more clearly shows that matrix deformation (voiding) occurs only close to the crack surface. This limited amount of matrix yielding leads to poor toughness. Note, however, matrix yield can be extensive if the test speed is reduced, whereby even the non-reactive blends are tough in slow rate testing. Figure 7-Bottom shows one of many typical dilation bands found in a slowly loaded sample of the non-reactive blend. This band has extended several hundred microns away from the crack tip.

Figure8-Top, in contrast, shows the crack path of the Reactive-C sample. Again, there is no evidence of massive shear yielding in the matrix. Closer examination, however (Figure 8-Bottom), shows that extensive crazing has occurred in the matrix between these large well-adhered particles. It is this massive multiple crazing that accounts for the good toughness in this 20% impact modified blend, similar to the mechanisms active in impact modified polystyrenes. Such a mechanism can only occur around large well-bonded particle, whereby it is not activated by the large particles in the non-reactive blend. This crazing mechanism is also only active when the amount of impact modifier is sufficiently high, as discussed by Bucknall (13),

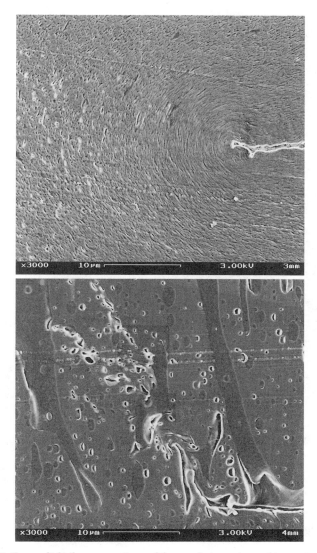

Figure 6. Cryopolished cross sections of the crack tip regions of impact modified PET. Top – 20% Reactive-B. Bottom – 20% non-reactive. All cracks proceed from right to left.

152

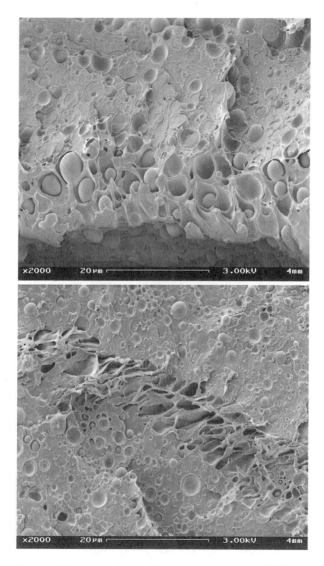

Figure 7. Freeze split cross sections of the crack tip regions of PET containing 20% non-reactive impact modifier. Top – an impact Charpy bar. Bottom – a typical dilatation band in a slow-bend Charpy bar. All cracks proceed from right to left.

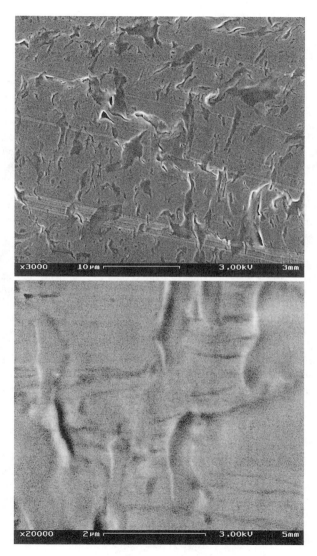

Figure 8. Cryopolished cross sections of the crack tip regions of impact modified PET. Top – Reactive-C. Bottom – crazes between impact modifier particles in Reactive-C. All cracks proceed from right to left.

whereby the 10% Reactive-A, B and C samples are brittle. This crazing mechanism, not accounted for in the interparticle distance model, allows PET to be tough over a very wide range of particle sizes.

Cavitation Model

In an attempt to rationalize the trends found in this system, the data were further analyzed using the cavitation model recently proposed by Bucknall (13-15). According to this model, toughness of a polymer blend is related to the interaction between the critical cavitation stress of the impact modifier, the shear and craze stresses of the matrix, and the applied stress state. Figure 9a shows a deformation diagram for an amorphous PET. The y-axis represents the effective stress, σe, as calculated by the Von Mises strain energy density criterion. The x-axis represents the mean stress, σm, calculated as $(\sigma 1 + \sigma 2 + \sigma 3)/3$. The line noted as "yield" depicts the pressure dependent yield strength for PET. The line noted as "craze" depicts the craze strength for the material, calculated as a constant $\sigma 1$. (Note these two curves were generated from data obtained on PET at impact speeds. The yield strength was measured using a tensile impact tester and the craze strength was measured from fractured 20 mil notched Charpy bars using slip-line theory. The yield strength obtained at impact speed is approximately twice the value measured at 50 mm/min.) The appropriate mode of failure can be predicted by following the appropriate loading line from the origin to the intersection with either the yield or craze curve. The lines shown as Bi, Iz and Ch represent loading curves for biaxial, Izod and Charpy loading, respectively. This plot accurately predicts that neat amorphous PET will yield under conditions of biaxial loading, but will fail in a brittle fashion under conditions of triaxial stress such as encountered in an Izod or high speed Charpy test.

The introduction of impact modifier into the PET reduces the yield strength of the blend, but the extent of that reduction depends on whether the particles have cavitated or not. Figure 9b shows the calculated yield curves for a blend containing 20% of non-cavitated particles (X) and cavitated particles (Z). The equations used to calculate these curves are described in greater detail in reference 13.

Figure 9c superimposes several hypothetical cavitation stress values for the impact modifiers used in this study onto the yield/craze envelope for the 20% impact modified blends. This model assumes that cavitation occurs at a constant value of mean stress. Cavitation strength increases with decreasing particle size (13), whereby the cavitation stress for the Reactive-C and non-reactive blend, with their larger particles, is placed at a lower value of mean stress than the cavitation stress for the Reactive-B blend. The relative placement of the Reactive-A sample will be discussed below.

Following the Izod and sharply notched Charpy load lines in Figure 9c reveals that the particles in the 20% Reactive-B blend will cavitate between the upper and lower yield envelopes. Upon cavitation, massive matrix shear yielding occurs in this blends, leading to good toughness under both Izod and Charpy conditions. In contrast, both the Reactive-C and non-reactive samples cavitate (or debond) in the

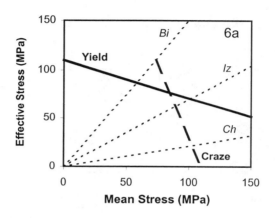

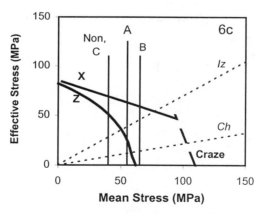

Figure 9. Cavitation diagrams for impact modified PET. See text for explanation

Continued on next page.

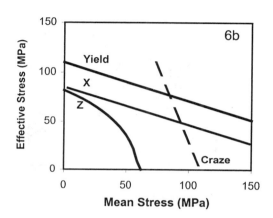

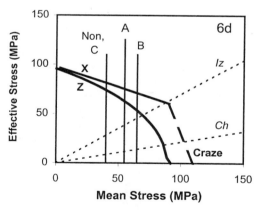

Figure 9. *Continued.*

elastic regime when loaded. This leads to premature craze formation. When the particles are not well bonded, as in the non-reactive blends, this crazing leads to premature fracture and poor toughness. In the blends containing 20% Reactive-C, however, the crazes are stabilized by the well bonded particles, and the blend is toughened by multiple crazing.

Of special interest is the use of this model to analyze the Reactive-A sample. Recall that this blend was tough under Izod loading conditions, but became brittle under the more constrained Charpy loading. According to interparticle distance theory this embrittlement occurred because the particles were too small to cavitate, and therefore general matrix crazing would have occurred. For this to have happened, however, the cavitation stress for this blend would have to be placed well above 100 MPa on Figure 9c, which is somewhat unreasonable. Furthermore, assuming that the cavitation stress does not change with test rate, this would imply that this blend would be brittle under slow loading conditions. (The measured craze stress is actually slightly lower in slow rate testing.) Instead, slow rate testing of sharply notched Charpy bars shows that this blend is quite ductile. A preferred explanation, suggested by the cavitation model, is that the cavitation stress for the Reactive-A particles is actually lower than that for the Reactive-B particles, even though the particles are smaller in Reactive-A than in Reactive-B. This would be quite possible if the Reactive-B were slightly crosslinked while the Reactive-A sample was not. With the cavitation stress for Reactive-A shown as in Figure 9a, the sample would shear yield under Izod conditions, but would cavitate prematurely under Charpy conditions. The small size of these particles would not support multiple crazing (13), whereby the Charpy sample would be brittle.

Figure 9d shows the cavitation diagram for the amorphous blends containing 10% impact modifier. The lower amount of impact modifier shifts the yield envelope of these blends to higher values of effective and mean stress than that shown in Figure 9c. Now, all cavitation occurs within the elastic regime, leading to craze formation from the particles, and brittle failure. At only a 10% loading, none of the crazes formed are stable, whereby none of the 10% loaded samples exhibited good toughness under either Izod or Charpy conditions.

Conclusions

The presence of two competing deformation mechanisms within the matrix of impact modified PET complicates any effort to rationalize toughness trends with particle size, volume fraction and adhesion of the impact modifier. In general, it was observed that supertough PET could be obtained by blending with 20% of a reactive impact modifier, such as E/MA/GMA. When this reactive impact modifier is compounded to produce a dispersed particle size of less than 200 nanometers (an interparticle distance of 50 nanometers), supertoughness is obtained through massive shear yielding of the matrix. When this reactive impact modifier is compounded to produce a dispersed particle size of 1000 nanometers, supertoughness is obtained through

158

multiple crazing. All samples that contained only 10% impact modifier were brittle. Supertoughness could be obtained using both amorphous and semicrystalline PET matrices. A cavitation model was found to be more useful in explaining the relative roles of volume fraction and particle size on toughness enhancement than an interparticle distance model. Nonetheless, a more rigorous study of the impact behavior of this system would be needed in order to further evaluate these models.

Literature Cited

1. *Rubber Toughened Engineering Plastics*; Collyer, A.A., Ed.; Chapman and Hall: London, 1994
2. Wu, S. *Polymer* **1985**, *26*, 1855
3. Wu, S. *J. Appl. Poly. Sci.* **1988**, *35*, 549
4. Wu, S. *Polymer International* **1992**, *29*, 229
5. Borggreve, R.J.M.; Gaymans, R.J.; Schuijer, J.; Ingen Housz, J.F *Polymer* **1987**, *28*, 1489
6. Borggreve, R.J.M.; Gaymans, R.J *Polymer* **1989**, *30*, 63
7. Kanai, H.; Sullivan, V.; Auerbach, A. *J. Appl. Poly. Sci.* **1994**, *53*, 527
8. Oshinski, A.J.; Keskkula, H.; Paul, D.R. *Polymer* **1992**, *33*, 268
9. Oshinski, A.J.; Keskkula, H.; Paul, D.R. *Polymer* **1992**, *33*, 284
10. Oshinski, A.J.; Keskkula, H.; Paul, D.R. *Polymer* **1996**, *37*, 4909
11. Argon, A.S.; Bartczak, Z.; Cohen, R.E. *Polym. Mater. Sci. Eng.* **1997**, *79*, 145
12. Muratoglu, O.K.; Argon, A.S.; Cohen, R.E.; Weinberg, M. *Polymer*, **1995**, *36*, 921
13. Bucknall, C.B. in *The Physics of Glassy Polymers, Second Edition*; Haward R.N; Young R.J., Eds; Chapman and Hall: London, 1997; p363
14. Lazzeri, A.; Bucknall, C.B. *J. Mater. Sci.* **1993**, *28*, 6799
15. Lazzeri, A.; Bucknall, C.B. *Polymer.* **1995**, *36*, 2895
16. Akkapeddi, M.K.; Van Buskirk, B.; Glans, J.H. in *Advances in Polymer Blends and Alloys Technology*; Finlayson, Ed.; Technomic Publishing Co.: Lancaster, PA, 1993; p 87
17. Stewart, M.E.; George, S.E.; Miller, R.L.; Paul, D.R. *Poly. Eng. Sci.* **1993**, *33*, 675
18. Epstein, B.N., U.S. Patent 4,172,859, 1979
19. Wang, X.H.; Zhang, H.X.; Wang, Z.G.; Jiang, B.Z. *Polymer* **1997**, *38*, 1569
20. Deyrup, E.J., U.S. Patent 4,753,980, 1988
21. Stewart, M.E., U.S. Patent 5,206,291, 1993
22. Akkapeddi, M.K.; Van Buskirk, B.; Mason, C.D.; Chung, S.S.; Swamikannu, X. *Poly. Eng. Sci.* **1995**, *35*, 72
23. Penco, M.; Pastorino, M.A.; Occhiello, E.; Garbassi, F.; Braglia, R.; Giannotta, G. *J. Appl. Poly. Sci.* **1995**, *57*, 329
24. Iida, H.; Kometani, K.; Yanagi, M, U.S. Patent 4,284,540, 1981
25. Hert, M. *Angew. Makromol. Chem.* **1992**, *196*, 89

Chapter 10

Yield Envelopes of Micro-Voided Epoxies

Alan J. Lesser, Robert S. Kody, and Emmett D. Crawford

Polymer Science and Engineering Department,
University of Massachusetts, Amherst, MA 01003

The macroscopic yield behavior of micro-voided epoxies is measured in uniaxial compression over a range of temperatures and void fractions. The results from these experiments agree well with current models predicting the yield behavior of porous materials. Additional experiments were conducted to measure the yield response of micro-voided materials over a range of stress states between uniaxial compression and biaxial tension. These results showed that the yield behavior of these micro-voided materials followed a modified von Mises yield criterion, which is in contrast to that predicted by a modified Gurson type model. However, no inelastic void growth was observed in these materials.

Introduction

It is well established that a critical micro-mechanism for effective toughening in many rubber-modified polymers involves the relief of hydrostatic stress through rubber particle cavitation or debonding, followed by inelastic matrix deformation.[1,2] Major forms of inelastic matrix deformation include void growth after cavitation, shear banding, and gross matrix deformation. However, the relative importance of particle cavitation on each of these mechanisms has been the subject of much controversy.

Some authors have suggested that a comparable toughness can be achieved in a polymer if the rubber particles are simply replaced by a pre-cavitated architecture; that is a micro-voided morphology. Their work has shown that polycarbonate with 20% voids by volume [3], and epoxies modified with hollow latex particles [4], pre-cavitated rubber particles [5], and non-adhering particles [6] have a significantly higher fracture toughness than comparable monolithic materials. These studies suggest that the cavitation process in itself may not significantly contribute to the energy absorbing mechanisms and that the primary role of the rubber particles is simply to nucleate micro-voids. These voids, in turn, relieve the hydrostatic stress and dissipate energy through inelastic matrix deformation.

Others believe that these findings are the exception rather than the rule, and that voids weaken and embrittle most glassy polymers [7,8]. With regard to this second view point, it has been proposed that rubber particles play a very important role in toughening of polymers, in addition to providing a microstructure more favorable for inelastic matrix deformation. Contradictory to this argument, Guild and Young have shown using finite element analysis that the stress concentration in the matrix material due to the presence of a void is quite similar to that of a rubber particle. Regardless of which viewpoint is correct, it is generally agreed that in this class of toughened polymers, the majority of energy dissipation occurs through inelastic matrix deformation.

The role of inelastic matrix deformation usually occurs through the formation of shear bands, dilatation bands, and inelastic void growth. Unfortunately, models that incorporate their contribution in the macroscopic yield response have been scarce in the literature. Until recently [9,10,12], the literature has been essentially absent of models that predict the yield response of rubber-modified polymers in arbitrary stress states, whereby cavitation and inelastic void growth are considered. In 1993 Lazzeri and Bucknall [10] made a first attempt at introducing a yield criterion for voided polymers. They extended a theory proposed by Gurson [11] for a perfectly plastic, von Mises material containing voids, to include the effect of hydrostatic stress on the polymer yield behavior by incorporating a coefficient of internal friction, μ_e. Lazzeri and Bucknall's yield function, Φ, was introduced as:

$$\Phi = \left(\frac{\sigma_e}{\sigma_0}\right)^2 + \frac{\mu_e \sigma_m}{\sigma_0}\left(2 - \frac{\mu_e \sigma_m}{\sigma_0}\right) + 2f\cosh\left(\frac{3\sigma_m}{2\sigma_0}\right) - f^2 - 1 = 0 \qquad (1)$$

where: σ_e is the von Mises equivalent stress

σ_0 is the yield stress in the absence of hydrostatic stress

σ_m is the hydrostatic stress

μ_e is the tensile equivalent coefficient of internal friction

f is the volume fraction of voids dispersed in the matrix.

In Equation 1, the yield envelope as described in a plot of σ_e versus σ_m of a cavitated or micro-voided morphology will be increasingly nonlinear with increasing void fractions. Steenbrink et. al [12] developed a yield model quite similar to Bucknall in that they also assumed a Gurson type response. Instead, however, Steenbrink et. al. ignored the pressure dependence of yield but incorporated factors that accommodate for large elastic deformation characteristic in polymers. Both equations predict similar relationships in practice (i.e., a nonlinear relationship in the predicted yield behavior as the hydrostatic stress is increased).

Previous studies by the authors[13] on CTBN modified resins showed no significant nonlinear response when tested up to 21% by volume modifier. These resins were tested in stress states ranging between uniaxial compression and biaxial tension. The results from this study are given in Figure 1 for an unmodified epoxy, an epoxy modified with 10% CTBN, and two other resins modified 21% CTBN rubber modifier. The triangles and oval symbols for the 21% CTBN modified epoxies represent two different particle size distributions; the oval symbols are for a 1.4 μm average particle size and the triangles are for a 5.3 μm average particle size. The data in Figure 1 show how the addition of rubber decreases the yield stress uniformly and promotes yielding in the stress states that have higher contributions of hydrostatic stress rather than failing in a brittle fashion.

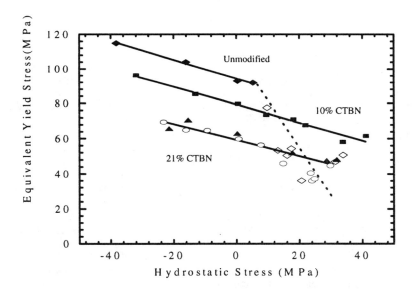

Figure 1: von Mises equivalent stress versus hydrostatic stress for rubber-modified epoxy. Data taken from reference 13. Solid symbols indicate samples that macroscopically yielded and hollow symbols indicate brittle failure.

Notice however, that none of the data presented in Figure 1 indicate the type of nonlinear response predicted by Equation 1. Subsequent microscopic investigations of the fracture surfaces also showed no significant post-cavitation inelastic void growth which could explain the discrepancies between Equation 1 and the data presented in Figure 1.

In this Chapter we present the results describing the yield envelopes on a micro-voided morphologies. The purpose of this investigation is to isolate the effects of inelastic deformation of the matrix in a pre-cavitated microstructure. The matrix material under investigation is a solvent-plasticized epoxy. Both void fraction and stress state are investigated. First we investigate how the void volume fraction affects compressive yield strength of epoxies. Next we characterize the yield/failure envelopes of epoxy networks with 4% and 28% volume fractions in stress states ranging between uniaxial compression and biaxial tension. The results from this investigation are then discussed in context of the cavitation viewpoints and models outlined in this introduction.

Experimental

Materials and Fabrication

The method used to make the micro-voided epoxies in this study follows closely to that recently reported by Kiefer, Hilborn and Hedrick [14], in which they chemically induced phase separation between a low molecular weight solvent and the epoxy before curing. Two types of voided epoxies were fabricated for this investigation: a hexane-modified epoxy and a propylbenzene-modified epoxy. The hexane-modified epoxy was made by reacting EPON 825 with a stoichiometric amount of 1-(2-Aminoethyl)piperazine (AEP), and the propylbenzene-modified epoxy was made by reacting EPON 828 with a stoichiometric amount of AEP. For both formulations, the resin was maintained at 45°C prior to mixing with the AEP. After mixing the resin with the AEP, premeasured amounts of solvent were added to the beaker. The resin, curing agent and solvent were then mixed together with a stir bar for several minutes. The mixture was degassed in a vacuum oven, and poured into molds for specimen fabrication. The curing of these materials typically involved three stages. Fabrication details for these materials are given in a reference [15].

In the first stage, the thin hollow cylinders are spun cast in an oven and the test hollow cylinders and plaques were placed in an oven to gel. The first stage is critical to development of the microstructure. The temperature at which the resins gel depends on their solvent concentration, the type of solvent, and the desired morphology. Typically for voided epoxies, larger quantities of solvent are added to the resin and the samples are gelled at relatively low temperatures. This causes the solvent to phase separate from the epoxy as the molecular weight of the network increases. At the end of stage one, the phase separated samples become opaque, as the phase separated solvent forms liquid spherical inclusions in the epoxy.

In the second cure stage, the crosslink density of the epoxy network continues to increase, but the cure temperature does not exceed the boiling temperature of the solvent. In the third and final post cure, the reaction between the epoxy resin and the AEP curing agent reaches full conversion. In order to make monolithic epoxy morphologies swelled with the solvents, the cure temperature was altered to suppress phase separation.

For the phase-separated materials, the cure temperature in the third stage was elevated above the boiling point of the solvent. This causes the solvent domains to evaporate, forcing the solvent into the epoxy matrix and also evaporating it out of the sample. This third stage leaves a voided epoxy material. It should be noted, however, that it was not possible to extract all of the solvent from the matrix material using this method. Typical morphologies of the micro-voided epoxies are shown in Figure 2.

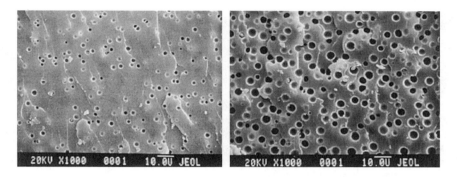

Figure 2: Scanning electron micrographs of micro-voided epoxies fabricated using low molecular weight solvents. Notice that both void fraction and size can be altered by altering processing conditions.

Sample Preparation and Testing

The resins were cast into compression bullets and thin-walled hollow cylinders for mechanical testing. All compression tests were conducted in accordance with the ASTM standard D695 at a strain rate of 0.1 mm/mm/min. The testing temperature was maintained in an environmental chamber and samples were conditioned for 30 minutes prior to testing.

The hollow cylinder tests were conducted on a biaxial tension-torsion machine, modified with an additional channel to supply internal pressure to the hollow cylinders. All tests were conducted in a pseudo strain controlled mode: with the axial strain controlled by crosshead displacement, the hoop strain controlled through internal pressure, and the radial strain calculated from constitutive equations. All samples were tested at a constant octahedral shear strain rate of 0.028 min^{-1}, and at temperatures noted in the figures and text. The hollow cylinders were tested in stress states ranging between uniaxial compression and biaxial tension. Full details of the hollow cylinder fabrication and testing procedure have been described in a previous paper [16].

Morphological Study

To examine the micro-voided morphology, samples were first cryo-fractured in liquid nitrogen. The samples were then coated with gold palladium and examined in a JEOL 35CF scanning electron microscope (SEM) in secondary electron imaging(SEI) mode. The average void sizes and concentrations were measured using quantitative stereology.

Results and Discussion

Effect of Void Concentration on Compressive Yield Strength

A series of compression samples were fabricated under different cure schedules with the propylbenzene-modified epoxy to yield a range of micro-voided morphologies. Two of these morphologies are shown in Figure 2. Their resulting void volume fractions and average void sizes were measured along with their glass transition temperature and compressive yield strengths. The results from these experiments are summarized in Table 1 below.

Table 1: Results on compression samples tested at 20°C and 30°C.

T_g (°C)	Void concentration (vol%)	Mean Void Diameter (μm)	σ_y^c @20°C (MPa)	σ_y^c @30°C (MPa)
49.5	1.7	1.7 ± 0.5	45.2	35.6
49.5	6.4	2.1 ± 0.3	43.9	37.0
48.3	7.7	2.4 ± 0.5	42.2	35.5
48.6	8.7	2.4 ± 0.4	41.0	34.3
53.0	13.9	3.2 ± 0.8	40.2	34.3
50.4	22.2	4.3 ± 1.2	36.8	32.8
52.5	24.2	4.2 ± 1.0	36.9	32.7
52.5	25.5	4.0 ± 1.0	37.2	31.5
52.0	26.5	5.3 ± 1.2	33.4	29.4
52.3	32.0	4.7 ± 1.2	34.1	28.2
52.8	32.4	4.4 ± 1.3	31.4	28.3

Note that the compressive yield strengths in Table 1 are convoluted by both differing amounts of residual solvent present in the epoxies, differing void volume fractions, and test temperatures. In order to isolate the effects due solely to void fraction, we first need to isolate the effects due to the residual solvent at each temperature.

Careful inspection of the data in Table 1 shows that increasing amounts of solvent produce greater reductions in the T_g of the resin. Earlier studies by the authors [16,17] on formulations that contained epoxies with different molecular architectures showed that the yield response of the materials collapse to a single value if the tests are conducted relative to a constant $(T - T_g)$ where T is the test temperature below the glass transition temperature. The results from Table 1 were plotted relative to a constant $(T - T_g)$ to determine if this model can describe the yield response of the epoxies swollen by solvent in this study. The compressive yield strengths of the non-voided epoxies that were swelled with varying amounts of propylbenzene are plotted in Figure 3 for a range of temperatures. From Figure 3, the decrease in yield strength with either an increase in test temperature and/or solvent content is apparent.

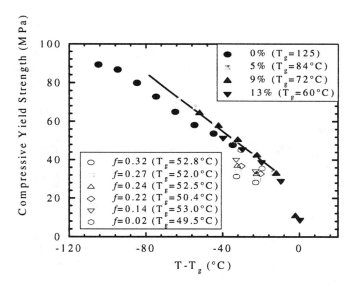

Figure 3: Compressive yield strength of non-voided and micro-voided solvent plasticized epoxies. The solid symbols are the non-voided materials and the hollow symbols represent the micro-voided materials.

In references 15 and 16, we showed that if the state of stress and strain rate are maintained constant, the yield strength, σ_y, of a series of aliphatic amine cured epoxies could be described by Equation 2:

$$\sigma_y = \sigma_y^{T_g} + \alpha\left(T - T_g\right) \text{ for } T < T_g \qquad (2)$$

where $\sigma_y^{T_g}$ represents the yield stress at T_g, and α is a negative term that describes the linear increase in σ_y as T is decreased below T_g. Equation (2) predicts that changing

T_g has the same effect on yield as changing temperature. In our previous study, we had changed T_g through a change in molecular weight between crosslinks, M_c. Figure 3 shows the compressive yield strengths from the swelled epoxies versus ΔT, the difference between the testing temperature, T, and T_g (i.e., $\Delta T = T-T_g$). In Figure 3, the compressive yield strengths of the swelled epoxies all collapse to a single curve, when compared at the same temperature shift below T_g, i.e. constant ΔT. This ΔT dependence is described by equation (2) and suggests that with regard to yield strength, the effect of changing T_g by swelling the epoxy is the same as changing temperature. There is a possible exception with regard to the pure epoxy, as the yield strengths of the pure epoxy tend to deviate from the others at lower temperatures. However, this deviation may be expected, as the pure epoxy has different intermolecular interactions from the swelled materials. In Figure 3, the solid line represents the experimental fit of the swelled epoxy yield data to equation (2), with $\sigma_y^{T_g} = 29.4$ MPa and $\alpha = -0.75$ MPa/°C. Note that only the yield strengths of the swelled epoxies were fit to equation (2). In Figure 3, the yield strength of several of the voided epoxies are also plotted, so that the yield strengths of the voided epoxies can be compared to the non-voided/swelled epoxies that were tested at the same temperature shift below T_g.

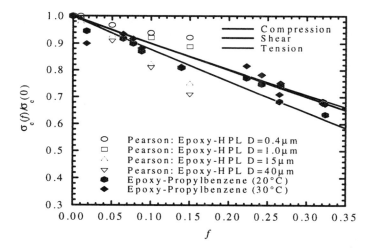

Figure 4: Normalized compressive yield strength versus void concentration for micro-voided epoxies. The solid data are the propylbenzene modified materials and the hollow data are hollow latex spheres taken from reference 4.

In Figure 4, the compressive yield strengths of the voided epoxies, normalized by the yield strength of the swelled epoxies tested at the same temperature shift below T_g, $\sigma_y(f)/\sigma_y(0)$, are plotted versus the volume fraction of voids. Figure 4 also shows how the reported compressive yield strength of epoxy is affected by the addition of hollow latex particles [4], as well as that predicted by Eq. 1 for uniaxial compression, pure shear and uniaxial tension.

Qualitatively speaking, the results in Figure 4 agree with many other reported results by others in that the compressive yield response is linearly related to the volume fraction of voids and/or rubber particles. In the Gurson model, both inelastic void growth and inelastic void collapse are kinematically admissible. This implies that Equation 1 should be valid in stress states that impose hydrostatic compression as well as hydrostatic tension provided that the voids do not completely collapse. Note that the reader should be cautioned in that inelastic void growth can evolve in an unbounded fashion while inelastic void collapse is bounded.

However, quantitatively speaking, Figure 4 shows that Eq. 1 slightly over-predicts the compressive yield strengths as a function of void concentration for the voided epoxies tested at 20°C and more seriously for the epoxies with hollow latex particles of 15μm and 40μm in diameter. The reader should note that Eq. 1 predicts a linear response between the yield stress and void volume fraction in the absence of hydrostatic stress (i.e., pure shear) and predicts a slightly nonlinear in all other stress states. Decreases in yield strength for the voided epoxies tested at 30°C and the epoxies with hollow latex particles of 0.4μm and 1.0μm in diameter are more in line with that predicted by Eq.1. However at 30°C the voided epoxies are approaching their T_g's, and hence equation (2) may not be as accurate at this higher temperature. As for the epoxies with the 0.4μm and 1.0μm hollow latex particles, the reported sizes of the particles included the hard shells of the particles. Therefore, we believe that the effective volume fraction of voids is lower than reported, because the hard-shells of the latex particles do not behave as voids.

Macroscopic Yield Envelope

Figure 5 summarizes the hollow cylinder results for the hexane-voided epoxy, where the von Mises equivalent yield strengths of each sample are plotted versus hydrostatic stress, σ_m. First it should be noted that the hexane-voided hollow cylinders were tested at 60°C to further suppress brittle failure and promote yielding. In Figure 5, the yield response shows no sign of nonlinearity as predicted by Eq. 1. We believe that these findings do not necessarily contradict the predictions of Eq. 1, as perhaps the 7% voids with an average void size of 6μm is most likely too low to experimentally isolate. Further post-mortem fractographic analyses of yielded samples showed no significant indication of inelastic void growth. Consequently, the combined effect of relatively large voids and low volume concentration may have contributed to the apparent discrepancies between that predicted in Eq. 1 and illustrated in Figure 5.

In an attempt to more thoroughly evaluate Eq. 1 ability to predict the yield response of micro-voided epoxies as the state of stress is altered, a second set of hollow cylinders were fabricated with the propylbenzene solvent. The propylbenzene modified cylinders contained a much higher void volume fraction (28%) and smaller

168

void sizes (average diameter of 3.5 μm). Figure 6 summarizes the hollow cylinder results for these propylbenzene-voided epoxies, where the von Mises equivalent yield and fracture strengths of each sample are plotted versus hydrostatic stress. The solid and hollow symbols represent macroscopic yield and brittle failure, respectively. In the propylbenzene-voided epoxies, there is again no observed nonlinearity in the relationship the von Mises yield stress and σ_m. However, Eq. 1 does predict significant nonlinearity at this concentration. Further, post-mortem fractographic studies on fractured specimens again showed no significant signs of inelastic void growth. These two observations suggest that the mechanism modeled in Eq. 1 is apparently not activated in the voided epoxies over the range of stress states they were tested. Although the lack of observable inelastic void growth may explain the discrepancies between the experimental results and that predicted by Eq.1, it does not explain why we were not able to produce experimentally observable inelastic void growth. This is the subject of further investigations ongoing in our labs.

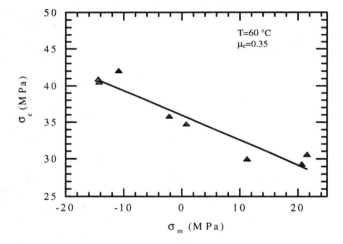

Figure 5: Plot of σ_e versus σ_m for the hexane modified epoxy formulation. The tests were conducted at 60°C.

It should be noted that only qualitative comparisons could be made between the behavior predicted by Eq. 1 and that measured on the hollow cylinders since the measured results are a combined consequence of the solvent plasticization, and void volume fraction. No attempt was made to deconvolute the solvent effects from the void fraction effects like was done for the solid cylinders in the uniaxial compression tests (see Figures 3 & 4) since this would require a significantly greater amount of experiments. Attempts were made to fabricate hollow cylinders under identical conditions to those used for the solid cylinders. Unfortunately, the spin casting process produced hollow cylinders with different residual solvent and void fractions from the solid cylinders produced by conventional castings.

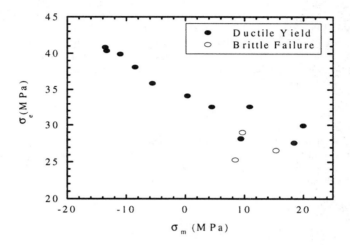

Figure 6: Plot of σ_e versus σ_m for the propylbenzene modified epoxy formulation. The solid symbols indicate macroscopic yield and the hollow symbols indicate macroscopically brittle response.

Conclusions

Micro-voided epoxies were fabricated with a variety of void sizes and volume fractions using low-temperature curing agents and low molecular weight solvents. The materials were tested in uniaxial compression using a solid cylindrical specimen geometry and over a range of stress states using a hollow cylinder geometry. Uniaxial compression tests on the solid cylindrical samples with void volume fractions ranging from 0 to 32% showed that the compressive yield strength of these materials decreases approximately linearly with increased void content, in a manner that is reasonably well predicted by modified Gurson yield function.

Hollow cylinders of these voided epoxies were tested in stress states ranging from uniaxial compression to biaxial tension. The results showed that the yield behavior of these micro-voided materials followed a modified von Mises yield criterion, which is in contrast to that predicted by a modified Gurson type model. However, in our studies no significant inelastic void growth was observed, which is the basis of the model. It is believed that the lack of inelastic void growth in the samples coupled with the lack of nonlinearity measured in the yield envelopes could be due to the range of void sizes and concentrations studied. That is, the void sizes produced via this approach are somewhat large compared to conventional rubber modified systems. In this context, the void sizes may be approaching the Griffith's flaw size for these materials thereby acting as defects. Another possibility, may be due to the fact that the residual solvent retained in the micro-voided epoxies dilate the resin network thereby kinematically reducing the deformational strain before failure. And finally, a third possibility may be that biaxial tension does not contain a sufficient enough dilatational component to promote significant inelastic void growth.

Acknowledgements

The authors wish to acknowledge the financial support from the Shell Chemical Co., Elastomers Department. The authors would also like to thank Professors Clive Bucknall and Andrea Lazzeri and Dr.'s Mike Modic and Mike Masse at Shell Chemical for their comments and insight in these studies.

References

1. Parker, D. S., Huang H. J., Yee, A. F., *Polymer*, **1990**,*31*, 2267
2. Dompas, D., Groeninckx, G., *Polymer*, **1994**, *35*, 4743
3. Hobbs, S. Y., *J. Appl. Phys.*, **1997**, *48*, 4052
4. R. Bagheri, R., Pearson, R. A., *Polymer*, **1996**, *37*, 4529
5. Y. Huang and A. J. Kinloch, *Polymer*, *33*, 4868 (1992)
6. Sue, H. J., Yee, A. F., *Polymer*, **1992**, *33*, 4868
7. Guild, F. J., Young, R. J., *J. Mat. Sci.*, **1989**, *24*, 2454
8. *The Physics of Glassy Polymers,"* 2nd *ed.*; Haward, R. N., Young, R. J., Eds: Chapman & Hall, London, 1997, p 508
9. Lazzeri, A., Bucknall, C. B., *Polymer*, **1995**, 36, 2895
10. Lazzeri, A., Bucknall, C. B., *J. Mat. Sci.*, **1993**, *28*, 6799
11. Gurson, A. L., *J. Eng. Mat. Tech.*, *Trans. ASME*, **1977**, *99*, 2
12. Steenbrink, A. C., Van der Geissen, E., Wu, P. D., *J. Mech. Phys. Solids*,**1997**, *45*, 405
13. Lesser, A. J., Kody, R. S., *Polym. Comp.*,**1999**, *in press.*
14. Kiefer, J., Hillborn, J. G., Hedrick, J. L., *Polymer*, **1996**, *37*, 5715
15. Kody, R. S., Crawford, E. D., Lesser, A. J., *J. Appl. Polym. Sci.*, in review
16. Kody, R. S., Lesser, A. J., *J. Mat. Sci*, **1997**, *32*, 5637
17. Lesser, A. J., Kody, R. S., *J. Poly. Sci.: Part B: Poly. Phys.*, **1996**, *35*, 1611

Chapter 11

The Network Structure of Epoxy Systems and Its Relationship to Toughness and Toughenability

H.-J. Sue[1], P. M. Puckett[2], J. L. Bertram[2], and L. L. Walker[2]

[1]Department of Mechanical Engineering, Texas A&M University,
College Station, TX 77843–3123
[2]Epoxy Products Department, The Dow Chemical Company,
Freeport, TX 77541

Beginning in about 1980, epoxy system networks were developed via an Edisonian approach that formed tough high T_g polymers (e.g., CET, crosslinkable epoxy thermoplastic resins). The driving force for the innovation was the need for inherently tough resins that could provide a base for formulation of aerospace prepregs and adhesives. Classical approaches to toughening epoxies either involved addition of toughening agents (rubber or thermoplastic) or reducing network crosslink density. Both approaches had significant drawbacks in their effects on polymeric properties and/or processability. This inherently tough CET chemistry was modified and extended in about 1990, to make it more suitable for liquid molding applications (e.g. resin transfer molding, pultrusion, etc.). Even though hundreds of CET formulations were investigated, no systematic investigation of the structure property relationships for this resin type was undertaken until recently.

Model CET epoxy networks, with variations in crosslink density and monomer rigidity, were prepared to study how the network structure affects modulus, T_g, and toughness (toughenability). Diglycidyl ethers of bisphenol-A, tetrabromo-bisphenol A, and tetramethylbisphenol-A, along with the corresponding chain extenders, were chosen to study how monomer backbone rigidity and crosslink density affect physical and mechanical properties of epoxy polymers. The present study indicates that, as expected, the backbone rigidity of the epoxy network, not the crosslink density alone, will strongly influence the modulus and T_g of epoxy resins. Upon rubber toughening, it is found that the rigidity of the epoxy backbone and/or the nature of crosslinking agent utilized are critical to the toughenability of the polymer formed. That is, the well-known correlation between toughenability and the average molecular weight between crosslinks (M_c), which usually corresponds to the ductility of the epoxy resin, does not necessarily hold true when the nature of epoxy backbone molecular mobility is altered. The potential significance of the present findings for a better design of toughened thermosets for structural applications is discussed.

Introduction

A very high percentage of the materials used in fabrication of aerospace composites and adhesives have been made from epoxy resins. Formulators are constantly pushing to extend the limits of epoxy technology to generate higher T_g, tougher polymers, with improved modulus, and lower moisture absorption, while retaining all the processability of conventional epoxy systems. Development of inherently tough epoxy resins that could either be toughened further by formulation with rubbers, elastomers, or thermoplastics, or could be used in liquid molding applications without addition of second phase toughening agents was believed to be particularly desirable.

In the last 20 to 30 years, it has been demonstrated that low molecular weight epoxy resins, which usually exhibit good processability, when cured with multifunctional curing agents, are brittle and cannot be effectively toughened [1-5]. The use of these materials for many structural and electronic applications is limited by this lack of ductility. On the other hand, when epoxy monomer molecular weights are high, although they become much more toughenable, their processability, rigidity, and use temperature can be greatly reduced. As a result, they begin to lose their suitability for composites, adhesives, and coatings applications. Resin formulation choice for these applications becomes a continuing series of trade-offs between viscosity, modulus, cure time, T_g, moisture resistance, and toughness.

The common approach to generating a toughened epoxy resin has been addition of a rubber or thermoplastic. As early as the late 1960s, rubber modification was reported as an effective approach for toughening brittle epoxy resins [1]. Since then, many workers have developed an understanding of how rubber particles toughen epoxy systems. The size and type of rubber particle in the epoxy network [2-4], the curing agent and crosslink density of the epoxy polymer [5-7], and the effect of the curing schedule on rubber particle phase separation [6,7], all influence the extent of matrix toughening achieved in epoxies and other highly crosslinked thermoset systems. While progress was made in understanding the physics of rubber toughening and the mechanics of fracture and crack propagation, compromises brought about by rubber toughening were often found to be too big a compromise for many applications [2-7]. This was especially true as compromises that enhanced properties began to decrease the processability of resin formulations. This type of compromise ultimately effects the cost to manufacture, a trade-off that industry is very reluctant to make.

To avoid the drawbacks associated with rubber toughening, thermoplastic modifiers were utilized to replace rubber particles in toughening [8-12]. However, it is found that the toughening effect becomes significant only when phase inversion takes place or when the thermoplastic composition reaches approximately 30% or more by weight [10,12]. Dispersion of thermoplastic particles in epoxy is typically a difficult art. In addition, the processability of these systems in comparison to unmodified resins is normally more complex and the other physical and mechanical properties of the modified resin are compromised for the sake of matrix toughness. Even though thermoplastic modified epoxies are commercially available as prepreg and adhesive formulations, there is still room for significant progress in this area.

Even with all the limitations of thermoplastics and rubbers as toughening agents for epoxy, state-of-the-art prepreg materials (e.g. Toray T800/3900-2, Hexcel IM-7/8551-7 or 8552, and IM-7/977-2) have been developed which provide exceptionally good impact resistant composite systems when properly processed [13-18]. These damage tolerant composites typically possess a resin rich interlaminar region, which has been toughened by the addition of relatively large elastomeric or thermoplastic particles.

This type of tough composite micro-architecture based on second phase toughening agents cannot be used in applications when a liquid molding process forms the composite. The use of large particulate toughening agents is unworkable primarily because the fibers act as a filter, trapping particles as the resin flows through them and making a controlled interlaminar thickness impossible. Controlling the thickness of the resin rich interlaminar region is a large contributor to the overall composite toughness [19]. Furthermore, second phase toughening tends to increase resin viscosity to a level that makes fiber wet out and saturation through thick fibrous bundles prohibitively difficult.

An approach to generating impact resistant composites that does not use toughened resin formulations is the use of a three dimensional (3-D) preform [20-22]. The presence of fibers in a Z-axis orientation physically ties together the lamina and that suppresses failure via a delamination mode. Composite fabrication via liquid molding techniques (e.g. resin transfer molding, RTM) using 3-D preforms and even resins known to form a very brittle polymeric matrix have generated some remarkably damage tolerant composite systems as measured by CAI [23-27]. However, the triaxial stress states that are induced by resin shrinkage and coefficient of thermal expansion mismatch during elevated temperature curing in the interstitial resin rich regions of a X-Y-Z reinforced composite are almost always relieved by microcrack formation. Use of a ductile resin that will not microcrack in this type of stress field is the preferred solution for RTM.

Development of ductile epoxy systems that are inherently tough and can serve as the basis for high viscosity prepreg and adhesive formulations as well as much lower viscosity liquid molding formulations was begun in the early to mid 1980s. This chemistry does not require second phase toughening to produce a ductile crosslinked polymer with high fracture toughness, but the addition of rubber and/or elastomers as would be expected further increases toughness.

Crosslinkable Epoxy Thermoplastic Resins

The crosslinkable epoxy thermoplastic (CET) chemistry, developed by Bertram, et. al. [28], has made it possible to produce commercially viable high T_g, high modulus, and low viscosity epoxies for high performance structural applications. The chemical design was based on two fundamental concepts: (1) the utilization of low molecular weights epoxy monomers with stiff backbones or bulky side groups to achieve high modulus and high T_g polymers [28-31] and (2) the promotion of *in-situ* chain extension of the epoxy backbone during cure, which then leads to a lower crosslink density epoxy network with greater processability and excellent inherent toughness and toughenability. The measured toughness for many of these formulations was often comparable to that of engineering thermoplastics, and they were referred to as CET resins.

A driving force in the creation and evolution of the CET materials was the desire to produce a formulated epoxy that was exceptionally tough, but was more processable than a typical two phase toughened epoxy system. Second phase toughening can be problematic to fabrication processes for a variety of reasons including, high viscosity, poor wet out of fibers or fillers, and final matrix toughness that is dependent on the thermal history of the cure. Because the CET chemistry begins with low molecular weight monomers, it has none of these problems. The rate of cure in the system can be adjusted by altering the cure temperature without any significant effect on the properties of the finished matrix.

The first commercial CET resin (designated Prepreg Resin 1) introduced required the addition of a curing agent and was targeted at the prepreg and adhesive formulators market. This CET chemistry evolved over the next several years to become a fully formulated resin system that could be used directly by molders that were fabricating by either pultrusion or RTM. A representative material of this class is shown as RTM Resin 2.

While the starting materials used to make these two representative CET resins are different, the properties are remarkable not only for their similarity, but also for their excellent overall performance. Table 1 compares the properties of unreinforced polymers of Prepreg Resin 1 (cured with stoichiometric DDS, diamino-diphenylsulfone) and RTM Resin 2. The ductility of the CET resin translates well, forming a very tough composite structure without the need for addition of rubber or thermoplastic. Properties of unreinforced clear castings are provided for only two resins, but they are representative of properties observed for dozens of CET systems that were made and characterized. When RTM Resin 2 and similar formulations were converted to composites (55% fiber volume) using AS4-6K 5HS fabric excellent properties were obtained. Typical ambient properties were SBS of 70 MPa, Compression Strength of 725 MPa and, CAI Strength of 262 MPa. With the low equilibrium moisture absorption of these materials, typically about 1.5%, there is excellent retention of properties up to temperatures of 121°C.

The examination of processing characteristics for CET resins also showed some unexpected results. A typical epoxy system on cure will release 300-400 joules/gram of energy. However, a typical cure exotherm for a CET resin is only 150-180 joules/gram. With significantly less energy being liberated during cure, there is less residual thermal stress found in a composite part. Also, typical cure shrinkage

Table I. Unreinforced Mechanical Properties of Prepreg Resin 1 and RTM Resin 2.*

Test	Measurement	Prepreg Resin 1	RTM Resin 2
Tensile	Strength (MPa)	90	83
	Modulus (GPa)	3.07	3.03
	Elongation (%)	5.5	5
Flexural	Strength (MPa)	131	145
	Modulus(GPa)	3.10	3.28
Moisture Absorption	Weight (%)	1.4	1
Density	Resin (g/cc)	1.5	1.3
Thermal			
--DSC	Tg (^{o}C)	168	167
--DMA[dry/wet]	Tg (Tan δ, ^{o}C)	168/---	174/167
--DMA[dry/wet]	Tg (E"$_{max}$, ^{o}C)	160/---	165/155
CLTE	below Tg, ppm/^{o}C	59	61
Cure Shrinkage	density change (%)	<1	<1

* Both resin systems were cured 4 hours @150°C and 2 hours @ 200°C.

measured on a variety of CET resins is less than 1%. By comparison, most epoxy resins will shrink from 2-5 % during cure. Therefore, this type of CET chemistry produces crosslinked polymers that have less thermal and shrinkage stress than typical epoxy systems.

The curing agent(s) and catalyst used with the epoxy determine the "processing window" of the CET resin. The catalyst system used in CET resins is latent and requires a certain amount of time at elevated temperature to break down and produce the active catalyst. The higher the temperature the less time required to form the active catalyst. Figure 1 provides a graph that shows the rate of viscosity increase for RTM Resins 1 and 2 at 150°C. The flat section of the viscosity profile for RTM Resin 2 shows an induction period of approximately 100 minutes at 150°C. The latent catalyst and the multifunctional phenolic curing agents used in this epoxy formulation provide this high relative stability. RTM Resin 1 is formulated using a latent catalyst with both multifunctional phenolics and aromatic amine curing agents. Even though at 150°C a similar induction period would be anticipated, amines and phenolic materials catalyze the reaction of each other with epoxy resins and this leads to the rather rapid increase in viscosity with time.

CET resins demonstrate unique thermal and mechanical properties and excellent processability. Curing forms a ductile thermoset polymer that does not require second phase toughening agents to generate high performance composites. Therefore, a decision was made to study the fundamentals of this polymer network in order to understand the relationship between structure and properties for this class of resins.

We are presently involved in an effort to understand on a fundamental level how the polymeric T_g, modulus, and toughenability are affected by the network structure in epoxies. The initial focus is on investigation of structure-property relationships in three model epoxy systems: (a) diglycidyl ether of bisphenol-A (DGE-BPA) epoxy monomer coupled with bisphenol-A (BPA) as a chain extender, (b) diglycidyl ether of tetramethylbisphenol-A (DGE-TMBA) epoxy monomer coupled with tetramethylbisphenol-A (TMBA) as a chain extender, and (c) diglycidyl ether of tetrabromobisphenol-A (DGE-TBBA) epoxy monomer coupled with tetrabromobisphenol-A (TBBA) as a chain extender. For each model epoxy system, the epoxy monomer and chain extender ratio is altered to vary the crosslink density of the model epoxies (Equation 1). The tetra-functional sulfanilamide curing agent, which contains two amines with distinctly different reactivities, was utilized to promote both chain extension and crosslinking of the epoxy network during cure.

Preparation Of Model CET Resins

The epoxy curing chemistry for the model systems to be studied is based on that of CET chemistry [28]. The controlled chain extension of the epoxy is accomplished by reaction of the difunctional epoxy with a difunctional phenolic (bisphenol). An anionic initiator catalyst, such as an ammonium, phosphonium, or sulfonium salt, is utilized as a latent catalyst to control linear chain extension and then crosslinking of the epoxy polymeric network.

Highly pure, commercially available epoxy monomers and chain extenders of DGE-BPA (D.E.R.* 332 epoxy resin) with epoxy equivalent weight (E.E.W.) of 174,

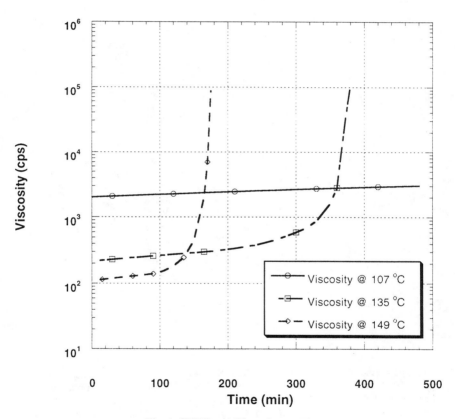

Fig. 1. CET Resin Viscosity vs. Time.

Equation 1

BPA, para-bisphenol A, DGE-TBBA (D.E.R.* 542 epoxy resin) with E.E.W. of 330, and TBBA (tetrabromobisphenol A) were obtained from Dow Chemical without further purification. TMBA (tetramethylbisphenol A) was obtained from General Electric and used as received. Epoxidization of TMBA into DGE-TMBA was performed in the laboratory using epichlorohydrin, isopropanol and distilled water. Excess epichlorohydrin and isopropanol were removed *via* vacuum distillation on a Roto-Evaporator up to a temperature of 150°C to achieve 98% yield of the epoxy product. The detailed experimental procedure was given in the patent literature [28].

The latent curing catalyst utilized for the present study was tetrabutyl phosphonium acetate-fluoroboric acid complex [28]. Reaction of the bisphenol and epoxy forms a product (see equation 2a,2b) that has a normal distribution of molecular weights around a mean that can be calculated based on the ratio of starting materials. The advancement of epoxy molecular weight is accomplished by phosphonium catalyzed reaction at 150-200°C until all phenolic functionality is consumed.

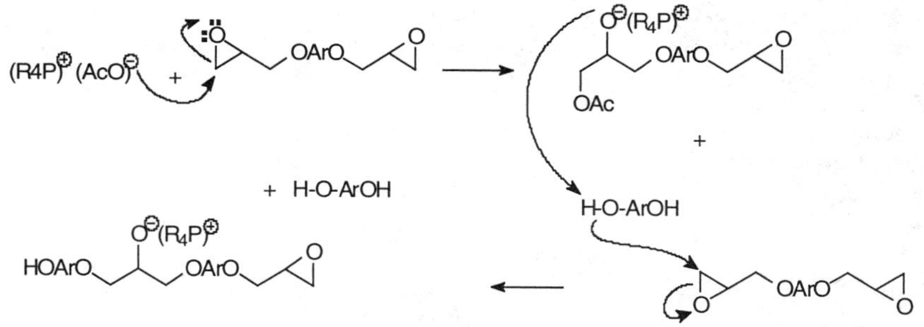

Equation 2a

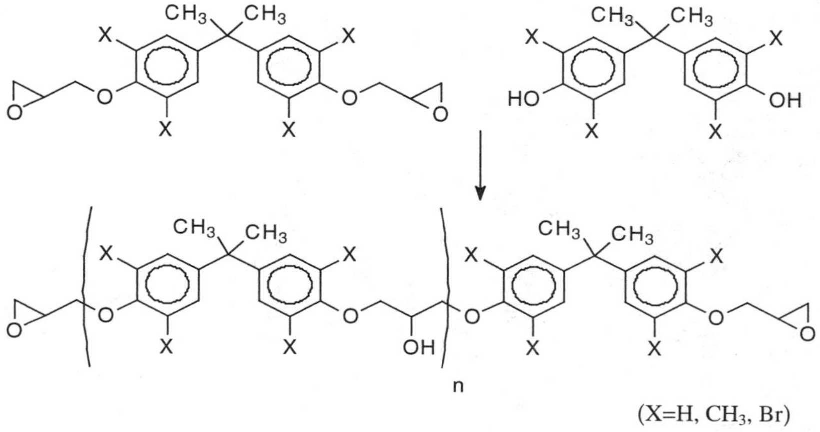

(X=H, CH$_3$, Br)

Equation 2b

Final cure of the epoxy system is accomplished by reaction with sulfanilamide. Sulfanilamide was chosen as a crosslinking agent because of its controllable difunctionality. The aniline end of the molecule is significantly more reactive than the sulfonamide. Therefore, adjusting the temperature of the cure controls the type of reaction expected based on the differential reaction kinetics of the amine and amide. Reaction occurs through the aniline nitrogen as shown in equation 3. The rate of this reaction is appreciably increased by the presence of a protic source to assist in the opening of the epoxy group. As always the more acid the proton source, the greater its effect on ring opening, and the faster the reaction rate. At 150°C no appreciable reaction occurs between the aniline end of the sulfanilamide and the epoxy until the bisphenol is added to the reactor (see equation 3). The sulfonamide is very sluggish and only reacts with the epoxy to generate the final crosslinked structure at elevated temperatures (equation 4).

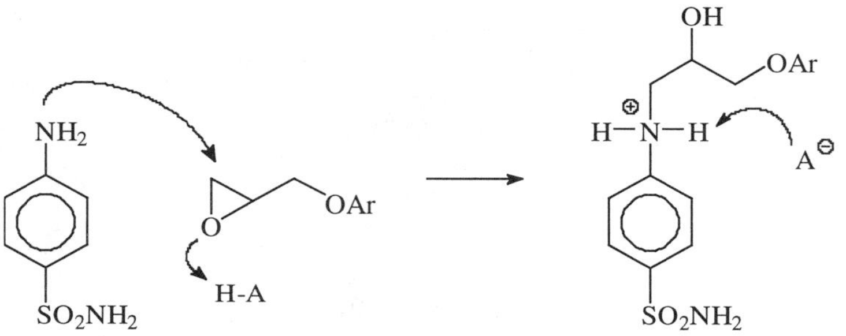

Equation 3

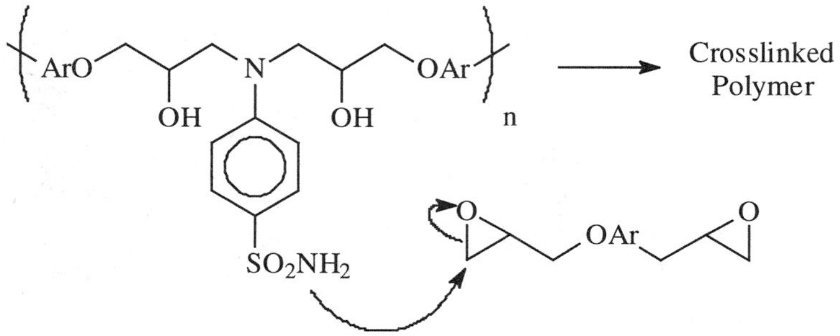

Equation 4

Three theoretical M_c values for each model epoxy were prepared. These three values of M_c are 600, 1,000, and 1,400. To achieve the designed M_c, the ratio of the bisphenol chain extender to the epoxy monomer is varied. The bisphenol chain extender and sulfanilamide must both be carefully accounted for when calculating stoichiometric ratios for the epoxy curative. To simplify the labeling, the M_c of 600, 1,000, and 1,400 represent the theoretical M_c of DGE-BPA model epoxy. For the DGE-BPA case the 600, 1000, and 1400 molecular weights represent approximately 2, 3, and 4 bisphenol groups between crosslinks respectively. The theoretical M_c for DGE-TMBA and DGE-TBBA are based on the same epoxy monomer and chain extender molar ratio as those of DGE-BPA. An underlying assumption is that the reactivity of BPA, TBBA, and TMBA chain extenders with an epoxy are about the same. As a result, the amount of –CH_3 and –Br present in the backbone will increase the actual M_c of DGE-TMBA and DGE-TBBA when compared with DGE-BPA. In other words, the M_c of DGE-BPA-600, DGE-TMBA-600 and DGE-TBBA-600 all have the same average distance between crosslinks, but differ in their weights by the number of –CH_3 or –Br groups which are used to replace –H in the backbone between crosslinks.

The epoxy resin, chain extender and sulfanilamide were mixed and heated until all components were dissolved, typically at about 150°C. Once dissolved, the mixture was cooled to 120-140°C and the catalyst added. After thoroughly mixing to ensure that the catalyst was homogeneously mixed, the mixture was vacuum degassed and poured into glass molds that were coated with a Teflon-based mold release agent. The molds were heated in an oven at 150°C for 4 hrs, then at 200°C for an additional 2 hrs. The molds were then cooled to ambient temperature in the oven and the castings de-molded.

Preparation Of Core-Shell Rubber-Modified CET Resins

Core-shell rubber (CSR) particles, which have a butadiene-styrene core (84 weight %) with a styrene-methylmethacrylate-acrylonitrile-glycidylmethacrylate shell (16 weight %) and have a uniform particle size of approximately 120 nm [32,33], were incorporated to study the toughenability of the model epoxies. Five weight percent of CSR particles was added to all the model epoxy resins, except for the DGE-TBBA system. The DGE-TBBA system has a vastly different solubility parameter from the other two model epoxy systems. As a result, the conditions were not found under which an acceptable dispersion of the CSR in DGE-TBBA could be obtained.

For toughened model CET epoxy systems, the CSR-modified model epoxy resins were prepared by mixing CSR with the epoxy resin component before adding chain extender, sulfanilamide and catalyst. The procedures for adding CSR into epoxy resins were reported earlier [33].

Test Specimen Preparations

Epoxy resin plaques, with and without CSR modification, with thicknesses of 0.635 cm (0.25") and 0.3175 cm (0.125") were cast. Samples with dimensions of 12.7 cm x 1.27 cm x 0.635 cm (5" x 0.5" x 0.25") were prepared for the double-notch four-point-bend (DN-4PB) [34-36] experiment. Specimens with dimensions of 6.35

cm x 1.27 cm x 0.635 cm (2.5" x 0.5" x 0.25") were cut for the single-edge-notch three-point-bend (SEN-3PB) plane strain critical stress intensity factor (K_{IC}) fracture toughness measurements [37,38]. These DN-4PB and SEN-3PB bars were notched with a milling tool (100 μm tip radius), followed by liquid nitrogen chilled razor blade tapping to wedge open a sharp crack with a parabolic crack front. The ratio between the final crack length (a) and the specimen width (W) was held in the range between 0.4 and 0.6.

Dynamic Mechanical Behavior

The dynamic mechanical behaviors of the neat and CSR-modified model epoxies, having dimensions of 5.08 cm x 1.27 cm x 0.3125 cm (2" x 0.5" x 0.125"), were studied using DMS (Rheometrics® RMS-805) under a torsional mode, with 5°C per step. A constant strain amplitude of 0.05% and a fixed frequency of 1 Hz were employed. The samples were analyzed at temperatures ranging from -150°C to 200°C. The temperature at which the primary Tan δ peak was located was recorded as T_g.

The dynamic mechanical spectra (DMS) of CSR-modified DGE-BPA and DGE-TMBA model epoxies and DGE-TBBA model epoxy systems are plotted in Figs. 2-4. Since it has been shown that the preformed CSR particles only slightly reduce shear storage moduli (G') below the T_g of the model epoxies if the CSR content is less than 10% by weight, the DMS plots of CSR-modified DGE-BPA and DGE-TMBA can be treated as neat DGE-BPA and DGE-TMBA systems for measuring T_g and rubbery plateau moduli [39,40].

The experimentally measured rubbery plateau shear moduli (Ge) and the equations proposed by Nielsen [41] or by Hill [42] can be utilized to estimate M_c of the model epoxies. The calculated M_c based on Nielsen's equation is given in Table II. The calculated M_c values are found to be in good qualitative agreement with the theoretical predictions for all three sets of model epoxies. The polymerization kinetics among the three sets of model epoxies appear to be the same as for low molecular weights model epoxies. That is, the M_c among the three sets of model epoxies appear to differ only in the molecular weights of -H, -CH3, and -Br in the epoxy backbones. However, the discrepancies between the calculated M_c and theoretical predictions become much larger for high M_c systems. At this moment, it is uncertain what is the reason(s) behind these discrepancies.

As anticipated, the higher the crosslink density the higher the T_g observed for the model epoxies. However, the magnitude of the Tan δ curve values at temperatures between the γ-relaxation peak and the β-relaxation peak (T_g) among the model epoxies shows that the higher M_c epoxies exhibit slightly lower Tan δ values than those of lower M_c epoxies. These observations imply that the high crosslink density epoxies are more capable of dissipating energy than the low crosslink density epoxies at temperatures above the γ-peak. These findings further imply that ductility and toughenability in epoxies cannot be simply correlated with the magnitude of the Tan δ curve. The physical nature and the scale of molecular motion(s) that are responsible for the molecular damping behavior are much more relevant to ductility and toughenability in polymers [43-46].

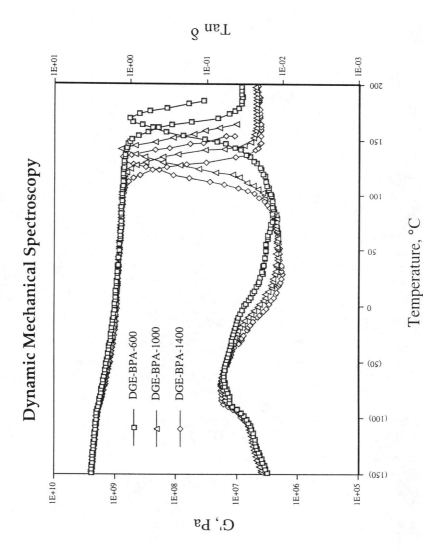

Fig. 2. DMS of DGE-BPA model epoxies. (Reproduced with permission from reference 55. Copyright 1999 John Wiley.)

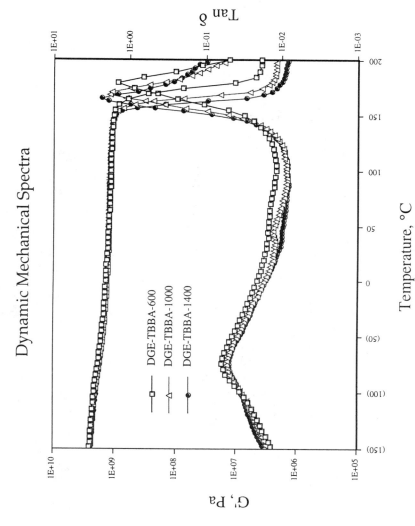

Fig. 3. DMS of DGE-TBBA model epoxies.

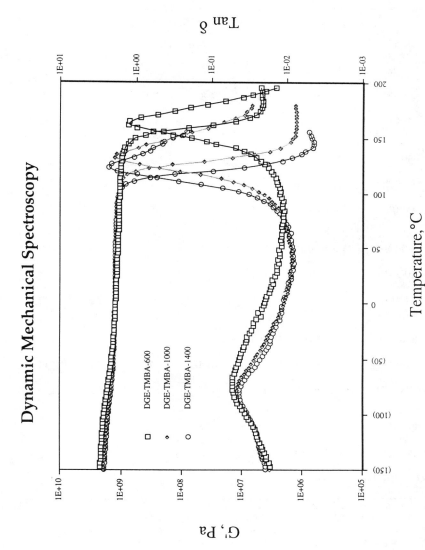

Fig. 4. DMS of DGE-TMBA model epoxies. (Reproduced with permission from reference 55. Copyright 1999 John Wiley.)

Table II. Mechanical Property of Neat and CSR-Modified Model Epoxies.

Resin Type	K_{IC} (MPa•m$^{0.5}$)	T_g (°C)	G'$_e$ (MPa)	M_C, est.[†] (g/mole)	E_f (GPa)	Density (g/cm^3)
600	1.00±0.12	170	8.6	390	3.03	1.233
1000	0.88±0.07	141	4.8	530	2.95	1.221
1400	1.04±0.07	135	4.1	580	2.85	1.214
600/R*	1.49±0.06	170	---	---	---	---
1000/R*	2.61±0.10	141	---	---	---	---
1400/R*	2.60±0.19**	135	---	---	---	---
600-Br	0.66± 0.10	180	4.0	930	3.79	---[§]
1000-Br	0.93±0.07	169	2.2	1640	3.72	---[§]
1400-Br	1.11±0.09	166	1.5	3530	3.72	1.872
600-CH$_3$	0.70±0.06	165	4.8	500	3.38	1.160
1000-CH$_3$	0.80±0.17	135	1.25	3460	3.45	1.145
1400-CH$_3$	0.52±0.15	123	0.73	6120[††]	3.62	1.139
600-CH$_3$/R*	2.02±0.20	165	---	---	---	---
1000-CH$_3$ /R*	1.69±0.19	135	---	---	---	---
1400-CH$_3$ /R*	1.25±0.09	123	---	---	---	---

*5% by weight of CSR is utilized for toughening.
**No longer valid for linear elastic fracture mechanics assumptions.
[†]Estimated using Log G' = 6.0 + 293ρ/M$_c$, where ρ: the density and G'$_e$: rubbery plateau modulus [41].
[††]Hill's approach is taken [42].
[§]Not measured; assumed to be 1.872.

For all three model epoxies, T_g is found to be inversely proportional to M_c as expected. However, the modulus dependence on M_c for the three model epoxies are much more complex and are significantly different from one another. The moduli of the DGE-BPA model epoxies are found inversely proportional to M_c. The moduli of the DGE-TBBA appear to be independent of M_c, and the moduli of the DGE-TMBA systems are proportional to M_c. The above findings strongly suggest that the molecular scale motion(s) that affect T_g and E of epoxies are quite different [47-50]. T_g appears to be affected by a larger scale molecular motion, which is restricted by the presence of crosslinks even when M_c is high. E appears to be sensitive to more localized molecular motions. Since -Br and -CH$_3$ functional groups have approximately the same size, it implies that the high density electron withdrawing –Br can induce additional restriction in molecular motion, which is much more effective in increasing E and T_g than the -CH$_3$ group.

The DMS results strongly suggest that, if the M_c estimations proposed by Nielsen and by Hill are correct for these systems, then the DGE-TMBA/TMBA model epoxy systems are significantly under-cured for monomer molecular weights above 1,000 g/mole. However, the epoxy titration, DSC, and FTIR probes suggest that the curing for the DGE-TMBA/TMBA model epoxy systems is every bit as complete as the other two model epoxies. The experimental validation of cure is more compelling than the theoretical calculations to the authors. However, additional work is needed to further substantiate these findings.

Fracture Toughness Measurements

A Sintech-2 screw driven mechanical testing machine was used to conduct the SEN-3PB experiments. Crosshead speeds of 0.0508 cm/min. (0.02"/min.) and 5.08 cm/min. (2"/min.) were utilized to conduct the SEN-3PB experiment for the neat epoxy resins and the CSR-modified model epoxy systems, respectively. The fracture toughness results are shown in Table II.

The CSR particles are shown to be effective in toughening the model epoxies even with only 5% by weight addition (Table II). The toughenability trend for the DGE-BPA model epoxies is as expected, i.e., the higher the Mc, the better the toughenability of the epoxy. The above finding has been reported by Yee and Pearson some years ago [5,6]. However, the DGE-BPA model epoxies exhibit a much higher toughness and T_g combination than the conventionally prepared DGE-BPA epoxies with similar M_c due to the unique epoxy curing based on CET chemistry [5,6,39]. The main reason(s) for such differences is believed to be due to the uniformity of crosslinking in CET networks and/or the use of sulfanilamide curing agent.

Surprisingly, for the DGE-TMBA model epoxy systems, the toughenability and M_c relationship is found to be the reverse of that for DEG-BPA systems. That is, the higher the M_c, the less toughenable is the epoxy. We surmise that since the sulfanilamide curing agent, when reacted with the epoxide groups, forms a rather flexible backbone segments with two three-carbon chains attached to the amine groups versus the short, "glycerine-like" segment produced by epoxy-phenoxy advancement. As a result, the more crosslinked portion of the polymer generates more flexible segments than the epoxy backbone. Therefore, the more ductile is the DGE-TMBA model epoxy.

It is noted that epoxies with high M_c are not necessarily more ductile than low M_c epoxies. That is, it may be more appropriate to use ductility, instead of M_c, to correlate with toughenability of epoxies.

Fracture Mechanisms Investigation

A Sintech-2 screw driven mechanical testing machine was used to conduct the DN-4PB experiments. A crosshead speed of 5.08 cm/min. was used for the CSR-modified epoxy systems. The fracture mechanisms in model epoxies were investigated based on the survived, subcritically propagated crack of the DN-4PB specimens. The detailed sample preparation procedures for optical microscopy (OM) and transmission electron microscopy (TEM) observations of the fracture mechanisms in this study are similar to earlier work and can be found elsewhere [2,8,33].

The toughening mechanisms found in neat and CSR-modified DGE-BPA model epoxies are similar to the neat and CSR-modified CET systems studied earlier [51]. They will not be elaborated here. For the DGE-TBBA model epoxies, only the neat resin was prepared. As a result, no detailed microscopic damage mechanism investigation was conducted.

For CSR-modified DGE-TMBA model epoxies, both OM and TEM were utilized to study the toughening mechanisms. Microtomed thick sections (about 1 μm) were prepared using a Reichert-Jung Ultracut E microtome. Since the sections are too thin to bring enough contrast for bright field imaging, only imaging under reflected light crossed-polars was performed. OM analysis indicates that the size of the birefringent plastic zone around the DN-4PB crack tip is largest for the DGE-TMBA-600 system (Fig. 5a), followed by DGE-TMBA-1000 (Fig. 5b), and by DGE-TMBA-1400 (Fig. 5c). The toughening mechanisms observed in DGE-TMBA-600 and DGE-TMBA-1000 are similar to those of the CSR-modified CET resin system [51], which possess a crack tip shear yielded zone encompassed by a larger rubber particle cavitation zone. Nevertheless, for DGE-TMBA-1400, the toughening mechanisms appear to be dominated by multiple cracking/croiding [39]. Only limited birefringent bands were found. These finding are consistent with the corresponding fracture toughness values (Table II), i.e., the fracture toughness is the highest for the DGE-TMBA-600 system, followed by DGE-TMBA-1000, and by DGE-TMBA-1400. TEM analysis was further performed to investigate the detailed toughening mechanisms.

When the DN-4PB damage zone are studied using TEM, as shown in Figs. 6-8, a high level of CSR particle distortion can be found around the crack tip and the crack wake of the DGE-TMBA-600 system (Fig. 6). This high degree of CSR particle distortion signifies a high level of plastic deformation of the matrix near these highly deformed CSR particles. In addition, cavitated CSR particles are found broadly distributed around the shear yielded zone. For DGE-TMBA-1000 system, the toughening mechanisms are similar to those of the DGE-TMBA-600 system except that the scale of CSR particle distortion is not as high (Fig. 7). In the case of the lowest M_c DGE-TMBA-1400 system, the crack tip damage zone appears to be limited

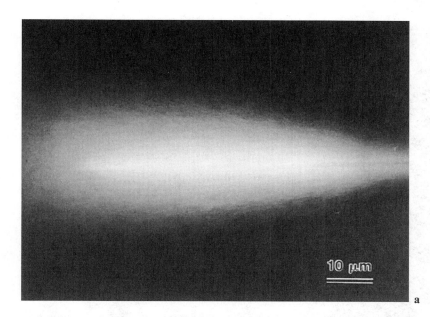

a

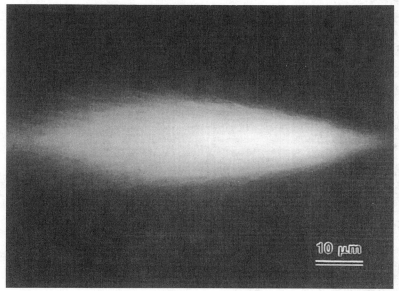

b

Fig. 5. Reflected optical micrographs of model DGE-TMBA epoxies under crossed-polars. (a) DGE-TMBA-600, (b) DGE-TMBA-1000, and (c) DGE-TMBA-1400. It is evident that DGE-TMBA-600 has the largest birefringent plastic zone, followed by DGE-TMBA-1000, and by DGE-TMBA-1400.

(Reproduced with permission from reference 55. Copyright 1999 John Wiley.)

Continued on next page.

c

Figure 5. *Continued.*

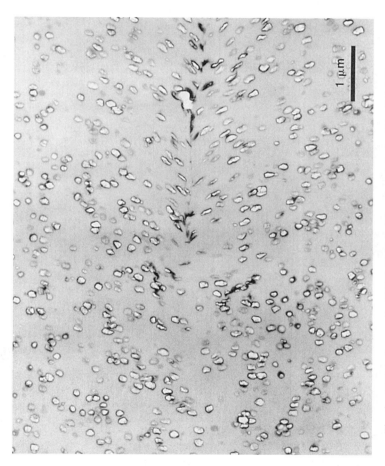

Fig. 6.   TEM micrograph of DGE-TMBA-600 at the crack tip of the DN-4PB damage zone.

(Reproduced with permission from reference 55.   Copyright 1999 John Wiley.)

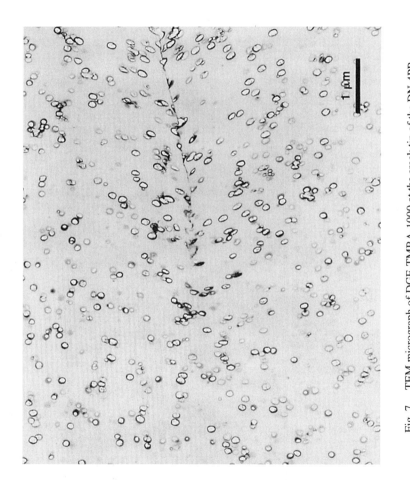

Fig. 7.   TEM micrograph of DGE-TMBA-1000 at the crack tip of the DN-4PB damage zone. (Reproduced with permission from reference 55.   Copyright 1999 John Wiley.)

to a few narrow, straight arrays of cavitated CSR particle bands (Fig. 8). These organized cavitated CSR bands were first observed by Sue [39] on CSR-modified DGE-BPA/piperidine system. In this scenario, since the dilatation bands formed around the crack tip are not as extensive as that of the DGE-BPA/piperidine epoxy, the toughening *via* dilatation bands is not as effective as those of the rubber particle cavitation and matrix shear banding mechanisms found in the model DGE-TMBA-600 and DGE-TMBA-1,000 systems. The above investigations are, again, consistent with the fracture toughness measurements and OM observations.

Figure 9 summarizes the relationship between fracture toughness and M_c for CSR-toughened DGE-BPA and DGE-TMBA systems. At this point, it is still unclear what causes DGE-TMBA to exhibit such an unusual toughness and M_c relationship, which is opposite to DGE-BPA system and to what is well known in the literature [3]. Additional work is currently underway to carefully probe the ductility, molecular scale motions, and network structures of these model CET resins. It is hoped that the underlying physics for the unusual toughness and M_c relationship can be revealed. It is also hoped that the source(s) for the formation of dilatation bands found in neat CET resins [52-54], which is believed to be caused either by network inhomogeneity or by certain characteristic molecular motions, can be identified.

The current research indicates that engineering thermosets with high T_g, high E, great processability, and good toughness/toughenability can be prepared. To achieve such desirable properties, it is necessary that we understand the curing chemistry as well as the nature of thermoset backbone mobility. At this moment, it is still unclear what causes the unusual M_c vs. toughenability relationship to occur in DGE-TMBA systems. However, it is certain that the CET technology is a plausible approach to produce commercially viable, tough thermosets for structural and electronic applications.

Conclusion

The toughening mechanisms in a series of model CET epoxy resins were investigated. It is found that T_g can be influenced both by crosslink density and by the backbone rigidity of the epoxy monomer. The backbone rigidity of the monomer also affect strongly the modulus of the epoxy. The toughenability of epoxy appears to be affected more by molecular mobility than by M_c. As a result, it is possible to design epoxy systems that exhibit high T_g, high E, and great toughenability and are easily processable by conventional methods.

Acknowledgment

The authors would like to thank E.I. Garcia-Meitin for performing TEM for this research. Financial support from The Dow Chemical Company is greatly appreciated. The authors also would like to thank the editor of the Journal of Polymer Science: Part B: Polymer Physics for his permission to reuse some micrographs and tables for this book chapter.

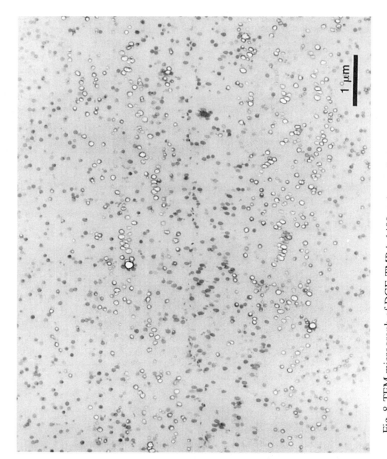

Fig. 8. TEM micrograph of DGE-TMBA-1400 at the crack tip of the DN-4PB damage zone. (Reproduced with permission from reference 55. Copyright 1999 John Wiley.)

M_c vs. Toughenability

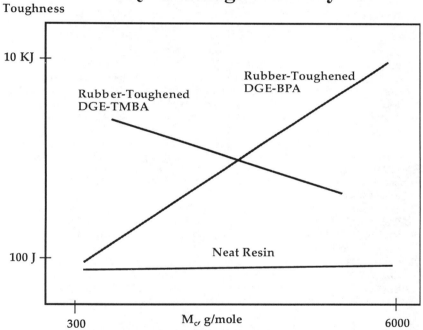

Fig. 9. A schematic showing the toughness and M_c relationship for DGE-BPA and DGE-TMBA systems.

References

1. Sultan, J.N., and McGarry, F.J., *J. Polym. Sci..*, 13, 29(1973).
2. Sue, H.-J., Garcia-Meitin, E.I., and Pickelman, D.M., "Toughening Concepts in High Performance Epoxies", Chapter 18 in Elastomer Technology Handbook, N.P. Cheremisinoff ed., CRC Press, Boca Raton, 1993.
3. Pearson, R.A. and Yee, A.F. *J. Mater. Sci.*, 24, 2571(1989).
4. G. Levita, Chap. 4 in "Rubber-Toughened Plastics", Adv. in Chem. Ser., 222, Ed. by C.K. Riew, p. 93, 1989.
5. A.F. Yee and R.A. Pearson, *J. Mater. Sci.*, 21, 2462(1986).
6. R.A. Pearson and A.F. Yee, *J. Mater. Sci.*, 21, 2475(1986).
7. Bradley, W. L., Schultz, W, Corleto, C., and Komatsu, S., "Toughened Plastics - Science & Engineering", Ed. by C.K. Riew, Adv. Chem. Ser., 233, p. 317, 1993.
8. Sue, H.-J., *Polym. Eng. Sci.*, 31, 275(1991).
9. Kinloch, A. J., "Rubber-Toughened Plastics", Ed. by C.K. Riew, Adv. Chem. Ser., 222, p. 67., 1989.
10. Bucknall, C. B., and Gilbert, A.H., *Polymer*, 30, 213(1989).
11. Pearson, R. A., and Yee, A. F., *Polymer*, 34, 3658(1993).
12. Pearson, R.A., and Yee, A.F., *J. App. Polym. Sci.*, 48, 1051(1993).
13. Bucknall, C.B., Chapter 4, "Approaches to Toughness Enhancement", Advanced Composites, Ed. Ivana K. Partridge, Elsevier Applied Science, NY, pp. 145-161, 1989.
14. Almen, G., Mackenzie, P., Malhotra, V., and Maskell, R., 35th International SAMPE Symposium, pp. 419-31, 1990.
15. Grande, D.H., Ilcewicz, L.B.,. Avery, W.B, and Bascom, W.D., 1st NASA Advanced Composites Technology Conference (NASA Conference Publication 3104), Seattle, WA, pp 455-475, 1990.
16. Lee, F.W., Boyle, M.A., and Lefebvre, P., 35th International SAMPE Symposium, pp 162-74, 1990.
17. Recker, H. G., Alstadt, V., Eberle, W., Folda, T., Gerth, D., Heckmann, W., Ittemann, P., Tesch, H., and Weber, T., SAMPE Journal, 26, 73(1990).
18. Recker, H.G., Alstadt, V., and Stangle, M., 37th International SAMPE Symposium, 493(1992).
19. Groleau, M.R., Shi, Y.-B., Yee, A.F., Bertram, J.L., Sue, H.-J., and Yang, P.C., *Comp. Sci. Tech.*, 56, 1223(1996).
20. Conference Proceedings: 3-D Composite Materials, Ed. H. Benson Dexter, E.T. Camponeschi, L. Peebles, NASA Conference Publication 2420, Annapolis, MD, 1985.
21. Funk, J.G., and Deaton, J.W., NASA Technical Paper 2590, 1989.
22. Brookstaiin, D.S. 35th International SAMPE Symposium, 746(1990).
23. Brosius, D., and Clarke, S., Advanced Composite Materials: New Developments and Applications Conference Proceedings, Detroit, MI, 1(1991).
24. Dow, M.B., Smith, D.L., and Lubowinski, S.J., NASA Conference Publications 3038, June 1989.
25. Palmer, R.J., Dow, M.B., and Smith, D.L., NASA Conference Publication 3104, pp 621-646.
26. Clancy, H.M., and Lufti, D.E., 18th International SAMPE Technical Conference, pp 135-141, 1986.

27. Vieth, W.R., Diffusion in and Through Polymer, Hanser Publishers/Oxford University Press, NY, 1991.
28. Bertram, J.L., Walker, L.L., Berman, J. R., and Clarke, J.A., US Patent 4,594,291, 1986.
29. Bauer, R., 18th International SAMPE Tech. Conf., Oct. 7-9 1986.
30. Dewhirst, K.C., US Patent 4,786,668, 1988.
31. Schultz, W.L., Portelli, G.B., Jordan, R.C., and Thompson, W.L., *Polym. Preprint*, 29, 136(1988).
32. Henton, D.E., Pickelman, D.M., Arends, C.B., Meyer, V.E., U.S. Patent 4,778,851, 1988.
33. Sue, H.-J., Garcia-Meitin, E.I., Pickelman, D.M., and Yang, P.C., "Toughened Plastics - Science and Engineering", Ed. by C.K. Riew, Adv. Chem. Ser., 233, p. 259, 1993.
34. Sue, H. J., Pearson, R. A., Parker, D. S., Huang, J., and Yee, A. F., *Polym. Preprint*, 29, 147(1988).
35. Sue, H.-J., *Polym. Eng. Sci.*, 31, 270(1991).
36. Sue, H.-J., and Yee, A.F., *J. Mater. Sci.*, 28, 2915(1993).
37. ASTM Standard, E399-90.
38. Towers, O.L., "Stress Intensity Factor, Compliance, and Elastic --Factors for Six Geometries", The Welding Institute, Cambridge, England, 1981.
39. Sue, H.-J., *J. Mater. Sci.*, 27, 3098(1992).
40. Sue, H.-J., Bertram, J.L., Garcia-Meitin, E.I., and Walker, L.L., *Colloid & Polym. Sci.*, 272, 456(1994).
41. Nielsen, L.E.J., *J. Macromol. Sci.*, C3, 69(1969).
42. Hill, L.W., *Progress in Organic Coatings*, 31, 235(1997).
43. Kishi, H., Shi YB, H, Y.B., Huang, J. and Yee, A.F., *J. Mater. Sci.*, 33, 3479(1998).
44. Liu, J.W., and Yee, A.F., *Macromolecules*, 31, 7865(1998).
45. Xiao, C.D., Jho, J.Y. and Yee, A.F., *Macromolecules*, 27, 2761(1994).
46. J.Y. Jho and A.F. Yee, *Macromolecules*, 24, 1905(1991).
47. LeMay, J.D., and Kelley, F.N., *Adv. Polym. Sci.*, 78, 115(1986).
48. DiMarzio, E.A., *J. Res., Nat. Bur. Stds.*, 68A, 611(1964).
49. Diamant, Y., Welner, S., and Katz, D., *Polymer*, 11, 498(1970).
50. Crawford, E., and Lesser, A.J., *J. Polym. Sci.*, Part B, Polym. Phys., 36, 1371(1998).
51. Sue, H.-J., Bertram, J.L., Garcia-Meitin, E.I., and Walker, L.L., *Colloid & Polym. Sci.*, 272, 456(1994).
52. Sue, H.-J., Puckett, P.M., Garcia-Meitin, E.I., and Bertram, J.L., *J. Polym. Sci. Polym. Phys. Ed.*, 33, 2003(1995).
53. Sue, H.-J., Yang, P.C., Garcia-Meitin, E.I., and Bishop, M.T., *J. Mater. Sci. Letters*, 12, 1463(1993).
54. Sue, H.-J., Garcia-Meitin, E.I., and Orchard, N.A., *J. Polym. Sci., Polym. Phys. Ed.*, 31, 595(1993).
55. Sue, H.-J.; Puckett, P. M.; Bertram, J. L.; Walker, L. L.; Garcia-Meitin, E. I. *J. Polym. Sci., Part B: Polym. Phys.* **1999,** *37,* 2137–2149.

Chapter 12

Polymethacrylate Toughened Epoxies

Alexandre A. Baidak, N. Tahir, and J. M. Liégeois

**Laboratoire Matériaux Polymères et Composites,
Université de Liège, Bât. B6, B–4000 Liège, Belgium**

A novel technique is proposed to toughen high Tg epoxy resins with high Tg polymethacrylate (PMA) by synthesizing the modifier in situ during the curing of the epoxy resin. This process shows no content limitation of modifier regarding the initial viscosity of the blend. Toughness of the DGEBA-DDS system can be more than doubled by introducing 30% wt of PMA. Still, a small Tg reduction is observed. The morphology is analyzed and its control is attempted by introducing glycidyl methacrylate or 2-hydroxyethylmethacrylate as comonomers which both act on the grafting between the phases. Miscibility of these systems as a function of the modifier molecular weight, was examined and it was found that low Mw modifiers should be avoided.

Introduction

Over the last three decades, toughening of thermosets and in particular of epoxies has been addressed in many different ways, providing new concepts and new materials. Starting with the rubber modified epoxies that showed surprisingly higher fracture resistance, but, at the same time, showing comprised thermomechanical properties, research quite naturally focused on other kinds of modifier. Bucknall and Partridge (1) were amongst the pioneers, introducing high-performance thermoplastic modified epoxy networks. These new materials revealed a distinct advantage over the elastomeric modified systems. Their fracture toughness could indeed be increased without sacrificing the heat distortion temperature or the modulus of such blends (2-11). Although these systems show outstanding properties, dissolving 15 % (wt) of thermoplastic in epoxy resin can have detrimental effects on the processing of impregnated reinforced clothes. Technical problems related to the viscosity of the system retarded the wetting of the fibers thereby resulted in porosity.

The viscosity penalty in conventional thermoplastic-modified thermosets has driven us to examine new possibilities. In the present study, an attempt was made to reinforce an epoxy resin with a high Tg thermoplastic that would be synthesized in-situ during the curing of the epoxy (*12*). Wetting rovings or cloths with a blend solution containing exclusively monomeric species should be easy to process. Transposing the concept of SIN (Simultaneous Interpenetrating Networks) in our process, we had to choose two reacting systems, one of which being the epoxy network precursors, the other being those of the thermoplastic modifier. DGEBA-DDS asserted itself quite naturally for the epoxy system. The modifier was selected amongst the highest possible Tg polymethacrylate: the polyisobornylmethacrylate (P-IBMA).

The two major requirements, namely, two different modes of polymerization for the SIN synthesis (a polycondensation and a radical polyaddition), and high Tg thermoplastic (reported values for the Tg of the polyisobornylmethacrylate range from 140 to 170 °C) are met.

Experimental

Table 1 gives the recipe of a typical composition. All chemicals were used as received without any further purification. The 4,4`-diaminodiphenyl sulfone (DDS) is first dissolved into the diglycidyl ether of bisphenol A (DGEBA) for 40 minutes at 120 °C with vigorous stirring. The methacrylic monomers were first passed through neutral aluminium oxide columns to remove the storage inhibitors. The precursors of the thermoplastic phase were then mixed and purged with dry nitrogen to remove the dissolved oxygen and then added to the epoxy. Unless otherwise specified, the thermoplastic content for the materials presented in this study is 20 % (wt). The mixture was then poured into oil heated Teflon coated moulds. The cure cycle was 1 hour at 120 °C followed by 5 hours at 180 °C. In all instances, the curing agent was used in stoechiometry with the total epoxy content assuming 4 reactive hydrogens in DDS.

Table 1: Components Description and Typical Recipe

Compound	Amount	Supplier
Epoxy network	**80 parts**	
Diglycidyl ether of Bisphenol A	75 % (wt)	Shell (Epikote 828 EL)
(epoxy group content 5360 mmol/kg)		
4,4' diamino diphenyl sulfone	25 % (wt)	Ciba (HT 976)
Methacrylate network	**20 parts**	
t-butyl peroxybenzoate	1 % (wt)	AKZO (Trigonox C)
1-dodecanethiol	0 - 2 % (wt)	Aldrich
Isobornyl methacrylate	Up to 100 %	Elf Atochem

Toughness was measured on razor-blade-notched samples using a Charpy three point bending test. Glass transition temperatures and G* were evaluated by DMTA

(0.1 Hz, 3 °C per minute), using a Rheometric 700. Morphology was examined with electron microscopy (SEM on fractured surfaces, TEM), and by optical microscopy on polished surfaces. The determination of thermoplastic modifier molecular weight had to be performed on the samples before gelation of the epoxy. Therefore, the cure reaction was stopped after the 120 °C plateau. Characterizing the Mw before the full cure is over has barely no influence, since at 180 °C, the radical polymerization should no longer occurs. The residual peroxide thermoinitiator is indeed totally decomposed long before reaching 180 °C. After one hour at 120 °C, the sample is dissolved in methanol, which is a non-solvent of the poly(isobornylmethacrylate), centrifuged and rinsed again several times with methanol until a neat white powder is obtained. The powder is then dried and dissolved in tetrahydrofuran for the GPC characterization. The Mw are expressed in PS equivalent Mw.

Results and Discussion

Effect of the modifier content

The first effect to examine is the influence of the modifier content on the properties of the SIN, and the fractured toughness is obviously of major interest (see Figure 1). A continuous increase in toughness is observed with increasing the thermoplastic content (at least up to 30 % (wt) of modifier). The original toughness of the pure epoxy (0.64 $MPa.m^{0.5}$) is more than doubled when adding 30 % (wt) of P-IBMA. To insure a thermosetting continuous phase, the modifier content was limited to 30 % wt. It must be underlined here that, regarding the experimental or technical procedure (e.g. the viscosity of the initial blend), the modification of the epoxy with as much as 30 % causes no limitation.

Thermomechanical analyses were performed to assess the purity of the two phases. The DMTA run on the 30 % modified sample shows that the glass transition of the epoxy rich phase (Tg_E) falls off from 198 to 172 °C. The Tg_M, glass transition of the modifier rich phase, is measured at 118 °C (much lower than the 140-170 °C reported in the literature). Each of these two unexpected values need to be discussed. In the first place, the reduction of the Tg_E may have various origins. Although a two phases system is observed (two Tg and see morphology discussion), part of the thermoplastic modifier might be dissolved in the epoxy phase, and hence plasticizing the network. Monomer residues might also act as plasticizer. As discussed later, the full conversion of the SIN is rarely observed, no matter the system. A last consideration suggests that, during phase separation, the thermoplastic phase would extract selectively one of the epoxy network precursor, leaving the reacting matrix off the stoechiometry. Methacrylate monomer residues, monomers or oligomers from the epoxy system dissolved in the polyisobornylmethacrylate are likely to explain the difference between the measured Tg_M (118 °C) and the expected value.

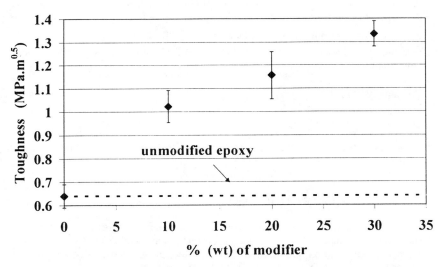

Figure 1: Influence of the modifier content on the toughness.

Controlling the Modifier Synthesis

Amongst the numerous parameters that control the morphology in a multi-phase reacting system, the miscibility between the various components is of prime importance. Controlling the miscibility is an indirect method to control the morphology.

The miscibility between two polymers is described by ΔG, the free energy of mixing. $\Delta G = \Delta H - T\Delta S$. There are two contributions, one of which being the enthalpic (ΔH), and the other, the entropic (ΔS) contribution. The previous expression can be written as

$$\Delta G = RT\left\{\frac{\phi_e \rho_e \ln(\phi_e)}{\overline{M}_e} + \frac{\phi_m \rho_m \ln(\phi_m)}{\overline{M}_m} + \frac{\chi \phi_e \phi_m}{v_{ref}}\right\}$$

where the subscript e refers to the epoxy matrix and m to the thermoplastic modifier, $\overline{M}$ is the molecular weight and ϕ is the volume fraction. Since ϕ is always smaller than one, the two first terms, which relate to the entropic contribution, are negative and thus favor the miscibility. It becomes evident that keeping $\overline{M}_m$ as small as possible will favor a better miscibility. The last term expresses the enthalpic contribution.

We can rewrite χ as

$$\chi = \frac{v_{ref}(\delta_p - \delta_m)^2}{RT}$$

It shows that the difference between the two solubility parameters must be minimized to increase the mixing.

The influence of the molecular weight

The control of the Mw of the modifier is accomplished by using a transfer agent: the 1-dodecanethiol (DDT). The use of various amounts of DDT for the modifier radical polymerization produced a thermoplastic phase whose Mw varied from 450,000 g/mol (0 % DDT) to 12,000 g/mol (2 % of the DDT).

The samples are analyzed by DMTA. The analysis of tan(δ) is quite relevant for the discussion of the results. The sample modified by the thermoplastic with the highest Mw (450,000 g/mol) shows two distinct α transitions corresponding to each polymer respectively. The transitions are 49 °C apart. By decreasing the modifier molecular weight, the difference between the two Tg decreases and eventually, for the sample TA20 (Mw 12,000 g/mol) vanishes (see Table 2).

The two last samples (TA10 and TA20) show a single broad glass transition. The values reported for each phase result from the mathematical deconvolution of the broad transition into two separate ones. This single broad transition suggests that the system is eventually miscible (i.e. the thermoplastic is dissolved into the epoxy matrix). Assuming the compete miscibility, a gross estimation of the Tg of the blend may be calculated with Fox's equation. For sample modified with 20 % (wt) in P-IBMA, we find 163 °C, which has to be compared to 154 °C found experimentally. However SEM analysis on the fractured surfaces as well as optical microscopy of the polished surfaces show clearly that the samples are two-phased. A partial miscibility of each phase in the other could possibly explain the inward shift of both glass transitions.

It is interesting to examine the mechanical properties of these samples. Rather surprisingly and except for small Mw reduction where one could possibly find a slight improvement, there is no beneficial effect in reducing the Mw on the sample toughness (see Figure 2). The purpose of reducing the Mw of the modifier was to slightly increase the miscibility so that phase separation would appears somewhat later leading to a secondary phase with smaller domains, hence more numerous.

Table 2: Samples Description and their Properties

	DDT (%)	HEMA (%)	K_{IC} $(+/- L *)$ $(MPa.m^{0.5})$	Tg_E (°C)	Tg_M (°C)	Mw (g/mol)
Control	0	0	0.644 (0.058)	198	/	
TA1	0.11	0	1.249 (0.044)	175	126	451000
TA2	0.22	0	1.283 (0.072)	172	120	332000
TA5	0.51	0	1.137 (0.099)	168	141	90000
TA10	1.00	0	1.169 (0.079)	152	150	27000
TA20	2.06	0	0.768 (0.05)	154	151	12000
HE0	0.2	0	1.157 (0.102)	173	131	
HE15	0.2	15	1.305 (0.092)	170	119	
HE22	0.2	22.5	1.297 (0.089)	171	112	
HE30	0.2	30	1.393 (0.118)	171	109	

* L is the confidence limit for a probability level of 95%

The effect observed for the very low Mw can be partly attributed to the Young modulus reduction. Dissolving a macromolecule in a reacting epoxy loosens the network, hence the Tg and the modulus.

However it must be kept in mind that the introduction of the DDT also modifies the kinetics of the radical polymerization .When a radical is captured by the transfer agent, the reinitiation is then happening with a defined rate that often differs from 100 %. This is the case with the 1-dodecanethiol. The first consequence is that the polymerization rate is then reduced. One hour at 120 °C might not be sufficient to fully polymerize the methacrylate, leaving unreacted monomers dissolved in the matrix.

At the present stage, it is rather difficult to predict the distribution of this unreacted monomer into the material. Considering that most of it would be dissolved

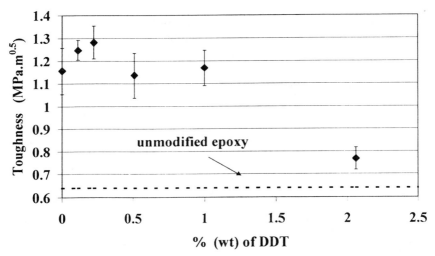

Figure 2: Influence of the transfer agent content on the toughness.

into the thermoplastic phase would appear rather natural (since like dissolves like). However, Sperling and Mishra (*13*) have addressed this question theoretically and experimentally for an polyurethane-polymethylmethacrylate interpenetrating polymer network. Based on a thermodynamic analysis, it appeared clearly that the entropic contribution was almost completely governing this problem, no matter how large was the difference in the solubility parameters between the monomer and the two networks. It resulted an even distribution between the two phases. Besides the thermodynamical approach, mechanical and sterical considerations interfere in some particular cases. These are for example, crystalline polymers or very densely crosslinked networks where the monomer is expelled out of one of the phases.

Influence of the Adhesion

To reinforce the interface between two polymers, one could choose to promote the physical interactions between the two types of macromolecules. Since a two-phase system is the desired morphology, increased interactions between the two polymers would eventually modify their miscibility and favor mixing. Another approach is to be tried: the chemical bonding between the phases. A grafting agent, namely the glycidyl methacrylate (GMA), was introduced in the recipe. Because GMA presents both a methacrylic double bond and an oxirane group, it is particularly suitable for grafting the modifier onto the epoxy network. When using this comonomer, the recipe was adjusted to preserve stoechiometry. The DGEBA amount was reduced according to the oxirane equivalent introduced by the GMA. This correction was proposed assuming that all oxirane groups from GMA would react with the epoxy hardener (DDS). The results of these investigations are shown in Table 3 and will be discussed for samples modified by 30 % (wt) of polymethacrylate.

Rather unexpectedly, it appears that the presence of GMA is detrimental to the fracture toughness. Even though no value could be measured for the sample with 35%, replacing 25 % of the IBMA by GMA reduces the K_{IC} by 15 %. An early conclusion would suggest that a strong interface is to be avoided, given the unfavorable influence of the grafting of the GMA.

Table 3: Influence of the grafting with GMA

% of GMA in the thermoplastic	K_{IC} $(MPa.m^{0.5})$	Tg_E $(°C)$	Tg_M $(°C)$
0	1.391	172	118
15	1.250	170	105
25	1.193	172	106
35	n. a.	171	106

Samples are modified with 30% wt P-IBMA.

The fairly constant values of the Tg_E presented in Table 3 indicate that the epoxy rich phase does not seem to be affected by the grafting agent. Depending on the final morphology of the sample (i.e. two or single phase), one could reasonably expect some variations in the network structure. Indeed, considering the case of a single phase network, the multiple GMA units carried by a methacrylate macromolecule are crosslinking the epoxy network in many places and in quite a different way than DGEBA does. On the other hand, if we consider a two phase system, there will be oxirane functional groups trapped within the dispersed domains thus unable to participate to the epoxy network. The disruptions of the network mentioned above appear then to be minor. Looking at the transition peak of the modifier, it can be seen that the presence of GMA lowers the transition by ca 13 °C, regardless of the GMA content. Here again, the results must be discussed in the light of the morphology of the samples. The initial drop can be explained rather naturally by the substantially lower glass transition of the P-GMA homopolymer (46 °C). Introduced as comonomer, GMA reduces the Tg of the copolymer. However when GMA is present at high concentration, Tg_M remains constant at a value of 106 °C, and this behavior has been confirmed by DSC measurements (results not shown here). At the same time, sub-inclusions appear clearly at the high GMA level. A sufficient but not necessary condition for Tg_M to remain constant could be compromised effects between copolymer Tg reduction on one hand, and the increasing influence of the interphase boundary regions as the extent of grafting increases on the other hand. In addition, one may believe that the morphology seen on Figure 3a is the result of the coalescence of particles such as those seen on Figure 3b, the latter providing thus more volume of boundary regions than the former. This argument would be supported by a somewhat larger transition peak observed by DMTA for the modifier.

The variation in the morphology on the SEM pictures is not quite evident. Given the very rough and chaotic structure of the fractured surface, it is not easy to point out any clear evolution . Fortunately, TEM images clarify greatly the discussion (Figure 3). As previously described the P-IBMA homopolymer produces a semi co-continuous structure. Increasing the amount of GMA in the copolymer modifies the morphology leading to a nodular dispersed phase. This tendency is already observed for sample with 15 % GMA and becomes evident when using 35 %. Another interesting fact is the presence of sub-inclusions for high amounts of grafting agent. These are assumed to be epoxy rich sub-domains and this observation supports the previous discussion. Because of the high GMA content, when the phase separation occurs, some of the epoxy phase, which is chemically linked to the thermoplastic macromolecules, is dragged into the thermoplastic. These nodules do not tend to coalesce since there are now bonded at their interface.

It could be advantageous to delay the grafting later during the curing cycle. An early grafting onto the growing epoxy network traps the polymethacrylate molecules in the matrix, once the epoxy has reached the gel point. Whole or part of the phase separation of the thermoplastic would be so hindered. Resulting morphology is a partly dissolved secondary phase and the loss of the thermomechanical properties of

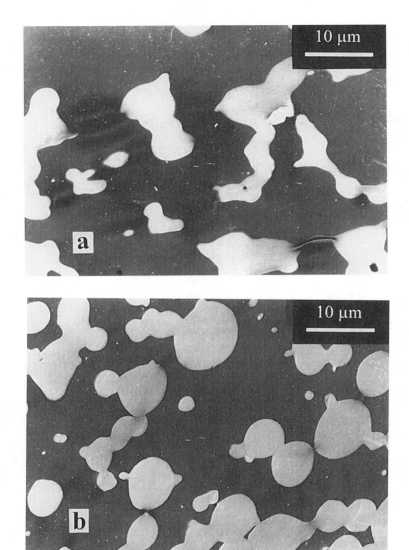

Figure 3: TEM images of 30 % (wt) modified samples showing the influence of GMA; with 0 % GMA (a), with 15 % GMA (b) and with 35 % GMA (c).

Continued on next page.

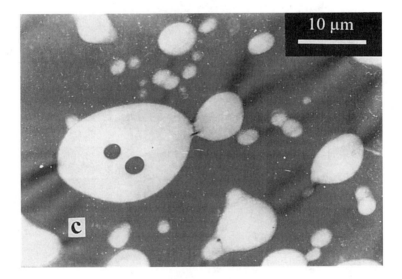

Figure 3. *Continued.*

the material due to the loosened network and the lower Tg of the modifier. Being so established, the toughening effect of such a morphology is nonexistent. The grafting between the two phases should only occur after the morphology is already defined and stabilised (i.e. in the later stage of the curing), hence after the complete *in situ* synthesis of the thermoplastic and its phase separation from the reacting epoxy. The 2-hydroxyethylmethacrylate (HEMA) was thus chosen as comonomer having a functional group able to react at high temperatures only (180 °C) with the growing epoxy network. Hydroxyl groups are known to react with the oxirane group of the DGEBA (*14*). This reaction is slow compared to the reaction of the oxirane and the primary amine, but can be considered to occur with a non negligible rate at high temperatures. We thus assume that at 120 °C the so modified polymethacrylate remains unreactive towards the epoxy network. The study of the radical copolymerisation was first performed in solution (20 % wt in the methylethylketone). Using the terminal model of Mayo and Lewis, values of r_1 and r_2 (where 1: IBMA and 2: HEMA) were found to be respectively 0.31 and 3.57 which can be described as a nearly ideal copolymerization since the product of r_1 and r_2 is close to unity. However, it must be stressed that due to the fact that $r_2 > r_1$, we must expect a certain extent of dispersity into the thermoplastic composition. The predictions given by the equation below which is based on the terminal model, show that in the early stages of the radical polymerization, the weight ratio of HEMA in the thermoplastic (average: 15%) would be of 38%.

$$\frac{M}{M^0} = \left(\frac{f_1}{f_1^0}\right)^\alpha \left(\frac{f_2}{f_2^0}\right)^\beta \left(\frac{f_1^0 - \delta}{f_1 - \delta}\right)^\gamma$$

with

$$\alpha = \frac{r_2}{1 - r_2} \qquad \beta = \frac{r_1}{1 - r_1} \qquad \gamma = \frac{1 - r_1 r_2}{(1 - r_1)(1 - r_2)} \qquad \delta = \frac{1 - r_2}{2 - r_1 - r_2}$$

where M is the total number of moles of unreacted monomers, and f_i is the molar fraction of the component i.

Due to its highly hydrocarbonated side group, the P-IBMA is indeed a non polar polymer, characterized by a rather low solubility parameter 16.6 $(J/cm^3)^{0.5}$ (*15*). This value is to be compared with the somewhat higher value of the epoxy resin which approximate 21 $(J/cm^3)^{0.5}$ (*15*). The introduction of the 2-hydroxyethylmethacrylate as comonomer will reduce the difference between the two parameters, hence favoring the miscibility. The solubility parameter of HEMA is not accessible from the literature. However, it can be estimated by first computing the solubility parameter of the monomer, using Small's group contribution method, and then extrapolating its value to the polymer. The only correction performed for this last step is to consider the molecular volume reduction when considering the polymer. A value of 24.2 $(J/cm^3)^{0.5}$ is then computed for the P-HEMA. Although it is only an approximation, it becomes evident that beyond grafting, compatibilization has obviously to be taken into account in the discussion of the results.

Depending on the extent of the miscibility enhancement, the system may end up with a single phase or the phase separation could occur at higher epoxy conversion and thus when the global viscosity is higher. Indirect methods as well as morphology analysis will be used to check the validity of the above assumptions. The dissolution of whole or part of the thermoplastic modifier is rejected by analyzing the glass transition values of the epoxy network which show to be unaffected by the increasing amounts of HEMA.

The introduction of the 2-hydroxyethylmethacrylate into the copolymer has a second important consequence. The modification of the glass transition of the modifier. Since the reported data for the Tg of the P-HEMA homopolymer (which range from 55 to 86 °C), are far below that of the P-IBMA, we expect a significant reduction in the Tg of the copolymer. Indeed, the introduction of 30 % (wt) of HEMA in the copolymer reduces the Tg by as much as 23 °C (Figure 4). The variation of the Tg of the thermoplastic modifier also produces a slight effect on the adhesion between the two phases. The actual cure and, in the first place, the gelation of the epoxy occurs while the sample is at 180 °C . Therefore, the morphology is set and definitely fixed at high temperature. Upon cooling from 180 °C to room temperature, the epoxy phase remain in its glassy stage. On the contrary, the thermoplastic modifier with a Tg of ca 120 °C will undergo cooling while in its rubbery phase from 180 °C to 120 °C. It is important to consider the difference in the linear expansion coefficient of the polymer on both sides of its Tg . A common polymethacrylate (e.g. PMMA) has a linear expansion coefficient of $2.55 \ 10^{-4}$ m/m°C under its Tg and $5.75 \ 10^{-4}$ m/m°C above (15). This variation may cause internal stresses at the interface between the epoxy matrix and the modifier. In the case of an epoxy modified by pure PMMA, debonding of the PMMA particles is largely observed in the sample.

The morphology of the samples is discussed by analyzing optical micrographies of polished surfaces. The morphology of the HEMA free sample can be described as a semi co-continuous structure. The dispersed phase has indeed a non-regular structure characterized by an intermingled network of cylinders whose diameter approximate 5 μm.

The introduction of 15 % (wt) HEMA in the copolymer has a direct effect on the morphology. The dispersed phase is now characterized by a nodular morphology. Particle size is heterodispersed, ranging from 1 to 5 μm. Increasing the amount of HEMA to 22 % (wt) leads to a more complex morphology. A nodular dispersed phase is still present, but its particle size ranges from 1 up to 20 μm in diameter. Moreover, the larger particles exhibit a tertiary phase of small (around 1 μm) epoxy particles. In addition to those particles, large domains of what is thought to be the coalescence of several large particles are also visible. Epoxy sub-domains can also be seen.

The same pattern is observed when the copolymer contains 30 % (wt) of HEMA. In this case, particles have a typical size of 10 μm, and the salami structure of the morphology is even more visible. Epoxy sub-inclusions have a diameter of 5 μm.

From the data of Table 2, we can see that the introduction of HEMA is beneficial to the toughness of the samples. For the same total amount of modifier, the toughness further increases of about 20 %, when replacing 30 % (wt) of IBMA by HEMA. This

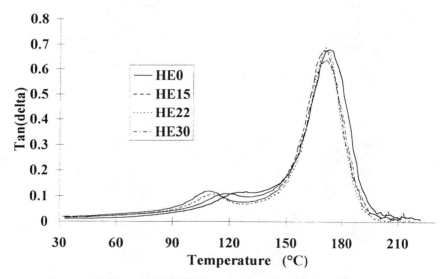

Figure 4: Influence of the HEMA on the dynamical mechanical properties.

beneficial effect cannot be attributed to the strength of the interface only, since we have also observed a change in the morphology of the samples. The increasing fraction of epoxy sub-inclusions in the phase separated thermoplastic acts as a pseudo-increase of the total amount of modifier if we consider its volume fraction.

Conclusions

The very brittle polyisobornylmethacrylate has proven to be efficient toughener for high Tg epoxies when synthesized during the curing of the resin. The elastic modulus is kept at a reasonably good level in the moderate temperature range. The role of the interface between the thermoplastic dispersed phase and the epoxy matrix is studied by copolymerizing the isobornylmethacrylate with two different grafting agents (glycidyl methacrylate and the hydoxyethylmethacrylate). A very strong interface such as given by the GMA appears not to be favorable to the toughness of the material whereas a moderate interface (using the HEMA) further increase the toughness compared to the P-IBMA homopolymer. However, the introduction of both comonomers modifies the morphology, hence the mechanical properties of the samples. Controlled miscibilisation is investigated by reducing the Mw of the thermoplastic. Low molecular weights (below 90.000 g/mol) dissolve excessively in the matrix, reducing its Tg and the toughness.

References

1. Bucknall, C. B.; Partridge, I. K. *Polymer* **1983**, 24, 639.
2. Raghava, R. S. *J. Polym. Sci., Part B, Polym. Phys.* **1988**, 26, 65.
3. *Toughened Plastics II;* Riew K.; Kinloch A. J., Eds.; Advances in Chemistry Series 252; American Chemical Society: Washington DC, 1996, p 405.
4. Bucknall, C. B.; Partridge, I. K. *Polym. Eng. Sci.* **1986**, 26, 54.
5. Bucknall, C. B.; Gilbert, A. H. *Polymer* **1989**, 30, 213.
6. Raghava, R. S. *J. Polym. Sci., Part B, Polym. Phys.* **1987**, 25, 1017.
7. Hedrick, J. L.; Yilor, I.; Wilkes, G. L.; McGrath, J. E. *Polym; Bull.* **1985**, 13, 201.
8. Kim, S. C.; Brown, H. R. *J. Mater. Sci.* **1987**, 22, 2589.
9. Cecere, J. A.; Hedrick, J. L.; McGrath, J. E. *SAMPE* **1986**, 31, 580.
10. Williams, R. J. J., Rozenberg, B. A., Pascault, J. P. *Adv. Polym. Sci.* **1997**, 128, p95.
11. Kim, B. S.; Chiba, T.; Inoue, T. *Polymer* **1993**, 34(13), 2809.
12. Tahir, N. Ph.D. thesis, Université de Liège, Liège, Belgium, 1999, in preparation.
13. Sperling, L. H.; Mishra, V. *Polym. Adv. Technol.* **1996**, 7(4), 197-208.
14. *Epoxy Resins Chemistry and Technology;* May C.; Tanaka Y., Eds.; Marcel Dekker Inc.: New York, NY, 1973.
15. *Polymer Handbook;* Brandrup, J.; Immergut, E.H., Eds; Wiley, New York, NY, 1989.

Chapter 13
The Effect of Particle Size on Synergistic Toughening of Boron Nitride–Rubber Hybrid Epoxy Composites

Michael F. DiBerardino[1] and Raymond A. Pearson[2, 3]

[1]IC Packaging Technology, Lucent Technologies,
555 Union Boulevard, Allentown, PA 18103
[2]Department of Materials Science and Engineering,
Lehigh University, Bethlehem, PA 18015–3195

The contribution of two modifiers, boron nitride (BN) and rubber to the overall fracture toughness of an epoxy polymer was determined as function of BN particle size and volume fraction. The use of both stiff inorganic and soft organic modifiers may promote the simultaneous occurrence of two or more toughening mechanisms potentially providing a synergistic increase in fracture toughness. Therefore, a systematic evaluation of BN/rubber/epoxy composites was performed using 10 and 50 μm diameter BN platelets. The use of 10 μm BN platelets and 2 μm rubber particles provided fracture toughness that is less than the additive contribution of the two mechanisms. Hybrid composites formulated with the larger BN particles exhibited fracture toughness values higher than the additive contribution of the two modifiers separately. Microscopic examination revealed that the larger BN particles were spaced further apart allowing for the development of the rubber-induced plastic zone in the epoxy matrix. If the BN particles are very close together, as is the case for the 10 μm BN platelets, the development of the rubber enhanced process zone may be suppressed. At the same volume fraction, larger particles will be spaced further apart and allow (and in some cases enhance) the formation of the rubber-induced plastic zone.

---

[3] Corresponding author

Introduction

Poor crack propagation resistance of epoxy polymers can be improved through two different approaches: 1) modification of the epoxy resin by the addition of compliant rubbery particles; and 2) modification of the epoxy resin through the addition of rigid inorganic fillers or thermoplastic particles. The addition of rubber particles to epoxies has been shown to initiate matrix plastic deformation which enhances the toughness, whereas rigid particles toughen epoxies through crack tip pinning, deflecting, crack wake bridging and microcracking. Experimental studies on both types of toughening have shown that the increase in fracture toughness reaches a maximum value at a particular volume fraction of the filler particles; i.e., the addition of particles beyond a critical volume fraction does not result in a further increase in fracture toughness in epoxies. (1-5). In order to provide additional fracture toughness to epoxy materials, hybrid-epoxy composites (epoxies containing both compliant elastomeric particles and rigid particles) have been investigated (6-11). The use of two modifiers in an epoxy polymer may promote the simultaneous occurrence of two or more toughening mechanisms. For hybrids this may result in an additional increase in fracture toughness above the toughness for a single particle type. If the two toughening mechanisms interact in a positive fashion, a synergistic increase in toughness may be achieved in that, for a given volume fraction of modifiers, the toughness of the hybrid would be greater than the additive contributions of the two modifiers separately. This synergistic toughening has been seen in epoxy systems containing rubber/ solid glass spheres (4,12), rubber/hollow glass spheres (5,7), and rubber/short fibers (8,11).

The contributions of multiple toughening mechanisms to the overall fracture toughness of a material have been addressed theoretically (13,14). A rubber-toughening model by Huang and Kinloch (14) considered the contributions of multiple mechanisms to the overall toughness of rubber modified epoxies. In their model, all three of the mechanisms present; rubber bridging, shear banding and void growth contribute in an additive fashion. That is, the overall toughness is the sum of the individual contributions. Conversely, a rubber toughening model by Evans et al. (13) predicts that process zone toughening mechanisms are additive while process zone mechanisms and crack wake mechanisms are multiplicitive. The occurrence of synergistic toughening in rubber modified epoxies has been studied experimentally by Pearson and Yee (15), Chen and Jan (16), and Riew et al. (17,18) using epoxies modified with a bimodal distribution of rubber particle sizes. These authors found that epoxies containing large and small rubber particles exhibited toughening that was greater than additive toughening as a result of enhanced shear banding between the larger rubber particles due to the presence of the small particles. In other words, the combination of large and small particles increases the amount of shear yielding of the matrix.

Prior Work in Hybrid Composites

Whereas rubber-rubber hybrids utilized two different sized rubber particles to toughen epoxy polymers, rubber-particulate hybrids incorporate rubber as well as rigid particles. The fracture properties of hybrid particulate composites containing rubber particles and a number of rigid particles such as glass spheres (7-10,19), silica (20) and alumina trihydrate (21) particles has been investigated by a number of authors. The majority of the work on rubber-particulate hybrids has been done using large 42 to 50 μm diameter glass spheres as the rigid particles. Kinloch et al. (6-8), Vallo et al. (22), Azimi et al. (9), and Smith et al. (10) all used large diameter glass spheres in conjunction with liquid rubbers (although Smith et al. (10) used 50 μm hollow glass spheres as opposed to the others that used solid glass spheres). Recently, Zhang and Berglund (19) used smaller 10 – 20 μm glass spheres.

Kinloch et al. (6-8) also investigated the effect of rubber particles and glass spheres on the fracture toughness of epoxies. Their formulations contained a fixed amount of rubber (15%), and varied the volume fraction of untreated and silane treated 50 μm glass spheres. The results of this study showed that the fracture toughness for the hybrid composites were higher than either of the formulations containing the individual modifiers. The increase in toughness is said to be a result of both crack pinning and matrix plastic deformation mechanisms being present. Additionally, a maximum in the fracture toughness data was observed for the hybrids at a glass volume fraction of 0.1. This maximum was present for both the untreated and silane treated glass spheres. The authors attribute the maximum to the debonding of the glass particles at higher volume fractions rather than any type of synergistic effect. The use of silane treated glass spheres produced a slight increased in fracture toughness over the hybrids with the untreated glass spheres. The authors believed this was due to higher pinning efficiency for the silane treated glass spheres.

This study, as well as those by Kinloch et al. (6-8) and Vallo et al. (22), has all shown beneficial effects of incorporating rubber and rigid particles. In some cases the authors even claimed to observe synergistic toughening. Unfortunately, these studies made no attempt to quantitatively assess the combination of the two toughening mechanisms.

Azimi et al. (9) and Smith et al. (10) attempted to assess the contributions of the rubber and rigid particles to the fracture toughness. These experiments were systematic studies in which the volume fraction of the rubber and the volume fraction of the rigid particles were varied while maintaining a total volume fraction of fillers at 10%. In this manner, the authors were able to compare the actual fracture toughness of the hybrid with the additive effect of the two particles separately. Synergistic toughening was seen for both rubber/solid glass sphere and rubber/hollow glass sphere formulations of 5%/5% and 7.5%/2.5%. Interestingly, the mechanisms associated with the rigid particles were not the same, Azimi et al. (9) observed crack pinning associated with the solid sphere whereas Smith et al. (10) observed microcracking and crack deflection associated with the hollow spheres. The nature of the synergism for both hybrids was found to be stretching of the plastic zone due to the presence of the glass particles. Azimi (9) attributed the plastic zone stretching to the interactions of elastic stress fields of the crack tip and the glass particles when the two are physically close together. Rubber particles engulfed in these overlapped stress fields are subjected to stresses high enough to promote cavitation/shear-yielding mechanisms. Optical micrographs from Kinloch et al. (7) show similar features. Additionally, Pearson and Yee (15) reported that small rubber particles in the vicinity of larger rubber particles cavitate extensively where as small particles that are rather far from the large particles do not.

Recently, Low et al. (11) reported synergistic toughening in hybrid-epoxy composites containing small rubber particles and aligned glassy metal ribbons or short alumina fibers. The authors quote that these hybrids are the toughest epoxy to date (K_{IC} of 4.2 MPa $\sqrt{m}$ for 15 phr rigid particles) due to the interactions between the rubber particle cavitation/matrix plastic deformation, and ribbon debonding, breakage, pull-out and bridging mechanisms (11).

Toughening Mechanisms

The addition of a rubbery phase can toughen the epoxy polymers by promoting process zone mechanisms such as rubber particle cavitation and concomitant matrix shear banding and plastic void growth (13-15). Shear banding and plastic void growth reduce the effective crack driving force by forming a plastic zone at the crack tip, thereby shielding the crack tip from the applied crack driving force. In general, for a given system, these mechanisms result in toughness that typically scales with the size of the plastic zone (15).

The addition of rigid particles to an epoxy polymer also can provide enhanced fracture resistance. Crack pinning may be the most commonly cited toughening mechanism for rigid inclusions in a brittle matrix. Evidence of crack pinning has been presented for rigid particulate systems containing glass spheres (2,7,9). In crack pinning, the moving crack front is impeded when it interacts with a rigid or impenetrable second phase material. The front is pinned at the location of the particles causing the crack to bow out between them. This change in the shape and length of the crack front will require more energy to propagate the crack.

If the rigid particles are in the shape of discs of rods, crack path deflection is a potential toughening mechanism. Crack tip deflection arises when interaction between the rigid particle and the crack produces a nonlinear crack path. Such a meandering crack causes an increase in the crack surface area as well as a decrease in the mode I driving force by deflecting the crack from the plane of maximum stress. The result of a deflected crack is an increase in the fracture toughness of the material (23) The presence of crack path deflection mechanism has been identified as the toughening mechanism present in BN modified epoxies (24,25).

Clearly there are benefits to using rubber particles as well as rigid particles to modify an epoxy resin. The use of rubber particles alone is a very efficient way to improve the fracture toughness of brittle polymers. However there is often a decrease in yield strength and stiffness associated with the addition of a compliant second phase material. The use of rigid particles is certainly not as efficient as rubber in toughening epoxy polymers, but there is no reduction in mechanical properties. In addition, the use of rigid fillers may provide desired physical properties such as thermal or electrical conductivity, and reduced thermal expansion. Both approaches to toughening have been shown to reach a limiting value for fracture resistance above which, adding additional does not result in a further increase in fracture toughness. The use of hybrid composites offers the potential to provide additional (and even synergistic toughening) along with desired physical and mechanical properties. Therefore, it is of great interest to study the nature of interactions among toughening mechanism induced by rubber particles and rigid inorganic particles.

The objective of this work is to understand the role of shear banding/plastic zone formation due to the presence of rubber particles with crack tip or crack wake toughening mechanisms associated with rigid particles. Clearly it is ultimately desirable to develop polymeric materials which contain rigid particles for the physical characteristics they provide and an elastomeric phase in which both particles contribute to the overall fracture resistance of the material. In the present work, a model composite system containing compliant rubbery and rigid inorganic boron nitride (BN) modifiers was prepared and the fracture behavior examined. The BN particles used were disc-shaped which presented an interesting challenge considering the majority of toughening mechanisms have been modeled for spherical or rod-shaped particles.

Experimental

The matrix resin used primarily in this investigation consists DER® 331 epoxy resin from the Dow Chemical Co. DER® 331 is a liquid diglycidyl ether of bisphenol A (DGEBA) with an epoxide equivalent weight of 187 g/eq. For investigations involving glass spheres, a higher molecular weight epoxy resin was used. DER® 661, a DGEBA epoxy with an epoxide equivalent weight of 530 g/eq was chosen. In all cases, the curing agent used was piperidine (PIP). The epoxy matrix has been modified by the incorporation of boron nitride platelets and rubber particles.

The boron nitride (BN) particles selected for this study were provided by Advanced Ceramics Corporation. Two BN powders were chosen, PT®-120 ($\approx$10 μm in diameter) and PT®-110 ($\approx$50 μm in diameter). The PT®-120 (10 μm BN) particles have a platelet or disc-shaped geometry with an aspect ratio (2r/t) of 10 and PT®-110 (50 μm BN) particles have an aspect ratio of 15. Figures 1 and 2 show SEM images of the BN powders.

To study the effect of hybrid composite blends, a rubber modifier was used in conjunction with the PT®-120BN and PT®-110 powders. The rubber modifier consist of a reactive liquid carboxyl terminated butadiene-acrylonitrile copolymer, Hycar® CTBN 1300 X8 (CTBN) from B. F. Goodrich. The CTBN particles were approximately 2 microns in diameter. The hybrid blends were formulated to have an overall volume fraction of modifier of 10%.

Plaques of the BN/rubber hybrid epoxy composites were prepared following a similar procedure to that use for the BN/epoxy blends. First 500 grams of DGEBA were heated to 80°C and degassed with agitation. Next the BN and rubber modifiers were added with mechanical mixing and allowed to thoroughly mixed for up to 4 hours. For the BN/CTBN hybrid, the particles and liquid rubber were added simultaneously to the neat epoxy. The resulting epoxy mixture was again degassed with agitation for at least 4 hours. Following this, 29 ml of piperidine were added to the mixture at ambient pressure with slow agitation. The final resin mixture was again degassed and poured into a 120°C pre-heated mold. The mold was then placed in a circulating air oven to cure for 16 hours at 120°C.

Fracture Toughness Evaluation

In this investigation, the linear elastic fracture mechanics approach is applied to measure fracture toughness. Fracture toughness is measured in terms of critical stress intensity measured in plane strain, K_{IC}. Fracture toughness values have been determined using pre-cracked, single edge notched (SEN) specimens in three point bending (3PB) geometry according to ASTM D5045-93 (26). Figure 3 contains a schematic diagram of the 3PB specimen geometry. Tests were performed using a screw-driven mechanical testing machine at a crosshead speed of 10 mm/min.

The following relationships are used to calculate K_{IC} from sample/pre-crack dimensions and sample deformation behavior:

$$K_{IC} = \frac{3PSa^{1/2}}{2tw^2} Y \tag{1}$$

Figure 1. SEM micrograph 10 μm BN powder showing disc-shaped particle geometry and a particle diameter of 10 μm and thickness of 1 μm.

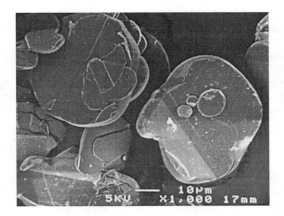

Figure 2. SEM micrograph 50 µm BN powder showing disc-shaped particle geometry and a particle diameter of approximately 50 µm and thickness of ≈3 µm.

where   P is the critical load for crack propagation (N);
                      S, the length of the span (mm);
                      a, the pre-crack length (mm);
                      t, the thickness (mm);
                      w, the width (mm)
                      Y, a non-dimensional shape factor given by where:

$$Y = 1.9 - 3.07\,(a/w) + 14.53\,(a/w)^2 - 25.11\,(a/w)^3 + 25.80\,(a/w)^4 \qquad (2)$$

K_{IC} values reported represents averages of a minimum of five tests.

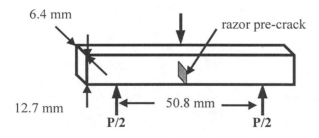

Figure 3. Schematic diagram of geometry of single-edge-notch (SEN) specimen tested in 3-point bend (3PB) configuration used for fracture toughness measurements.

Optical Microscopy

       Optical microscopy, OM, has been used to examine sub-surface damage in SEN fracture toughness specimens. To prepare unfractured specimens with a developed sub-surface damage zone, SEN-3PB specimens were loaded to approximately 90% of K_{IC}. OM samples were cut from the SEN-3PB specimen and mounted in a room temperature curing epoxy resin. To examine this damage, sections thin enough to transmit light are required. Typical sub-surface thin sections range from 40 to 100 microns in thickness. Thin sections were viewed using a Zeiss Axiomax light optical microscope under transmitted light (TOM) under both bright field and cross-polarized light.

Scanning Electron Microscopy

BN particles and SEN-3PB fracture surfaces have been examined using a JEOL 6300 scanning electron microscope. Samples were coated with a thin layer of chromium to prevent charge build-up on the sample surface. The conventional secondary electron imaging technique was used at 5kV accelerating voltage to obtain specimen and fracture surface images.

Results and Discussions

The results of the fracture toughness testing for the BN/rubber hybrid composites based on PT-120 (10 μm) particles are listed in Table I.

Table I. Fracture Toughness of 10 μm BN/Rubber Hybrids

Formulation	K_{IC} (MPa √m)	ΔK_{IC} (MPa √m)
Neat Epoxy	0.77±0.04	0.00
BN(5)	1.32±0.06	0.55
BN(10)	1.45±0.08	0.68
BN(5)/Rubber(5)	1.72±0.06	0.95
BN(2.5)/Rubber(7.5)	1.76±0.15	0.99
Rubber(10)	1.93±0.06	1.16

Note: Numbers in parenthesis indicate concentration of modifiers in vol. %.

These results show an increase in fracture toughness as the ratio of BN/rubber decreases (increasing rubber content). This is not surprising considering that the CTBN rubber is a more efficient toughening agent than the BN particles. To assess the contribution of each of the two modifiers and to determine if any synergistic toughening occurs, the data is plotted as a function of BN-to-rubber ratio in Figure 4. Unlike similar plots for hollow glass sphere/rubber (*9*) and solid glass sphere/rubber (*10*) hybrids the plot for BN/rubber hybrids does not show any indication of synergistic toughening effects. The BN/CTBN hybrids exhibit a fracture toughness that is nearly equal to the additive effect based on a rule-of-mixtures approach. However, unlike the data on glass/epoxies and rubber/epoxies, the increase in fracture toughness from the addition of BN particles is not linear in the range from 0 to 10%

as seen by the ΔK_{IC}, data in Table I. Thus the ROM approach in Figure 4, under predicts the additive effect of the two fillers. In order to present this data in a more appropriate manner Figures 4 also shows the sum of the individual contributions to the account for the non-linear K_{IC} behavior. It is evident comparing this data to the data on the neat epoxy, that the addition of BN to a rubber-modified epoxy provides an additional increase over that of the rubber alone but is in fact less than the additive contributions of the two particles. This indicates that there may be a competition between the crack path deflection associated with the BN particles and the rubber cavitation/shear yielding toughening mechanisms.

The use of transmission optical microscopy (TOM) to examine the crack tip should elucidate the interactions of the two toughening mechanisms. A TOM micrograph of the crack tip for the BN/rubber hybrids are shown in Figures 5 for the BN/CTBN systems containing 2.5% BN and 7.5% CTBN. What is most evident from this figure is the lack of a prominent cavitation or damage zone ahead of the crack tip associated with rubber cavitation. The TOM does display a very small, faint region of cavitation between the BN particles. The size of this cavitation zone is significantly smaller than the damage zone of an epoxy modified with rubber alone indicating that there is some degree of competition between two toughening mechanisms.

This behavior is noticeably different from that seen by Smith et al. (10) and Azimi (9) for hollow glass sphere/rubber and solid glass sphere/rubber hybrids. These authors reported synergistic toughening due to plastic zone stretching or branching as a result of the interactions between the rigid particles and the rubber particles. Azimi (9) attributed the plastic zone stretching to the interactions of elastic stress fields of the crack tip and the glass particles when the two are physically close together. Rubber particles engulfed in these overlapped stress fields are subjected to stresses high enough to promote cavitation/shear-yielding mechanisms. Optical micrographs from Kinloch et al. (7) show similar features. The BN/rubber hybrids exhibit a reduction in the plastic zone size rather than plastic zone stretching or branching as evidenced by the micrograph in Figure 5. This reduction in plastic zone size may be due to a number of factors that differ for previous systems that have demonstrated synergistic toughening, such as: resin ductility, size and shape of the rigid particles, as well as particle-matrix adhesion.

Effect of Inter-particle Distance on BN/Rubber Hybrids

Reviewing recent work on rigid particle/rubber hybrids reveals that the majority of the research on rigid particle/rubber hybrids has been performed on systems in which the rigid particles were 42-50 μm spheres. However, the BN/rubber hybrids studied to this point have used 10 μm disc shaped particles as the rigid filler. As a result of the size and shape of the BN particles, the BN/rubber hybrids blends with the same ratio as the glass sphere/rubber hybrids used by Smith et al. (10) and Azimi (9) will have significantly more rigid particles present in the hybrid system. The increased number of particles will force the distance between the particles to become very small. In order to reduce the number of BN particles and to increase the particles separation distance while still maintaining the same rigid filler/rubber ratios

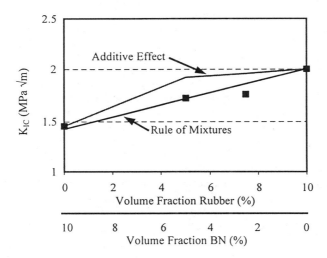

Figure 4. K$_{IC}$ versus modifier ratio for the 10 μm BN/CTBN rubber epoxy hybrids. Note the less than additive toughening for all the composite ratios.

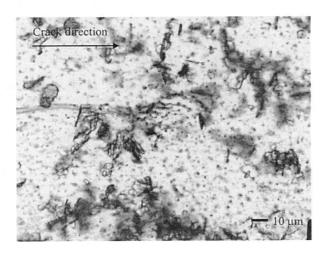

Figure 5. TOM micrographs for BN(2.5)/CTBN(7.5) hybrids showing competition between toughening mechanisms results in plastic zone compression.

that have been previously examined, hybrids were formulated using the larger (50 μm diameter) 50 μm BN particles. The results of the fracture toughness testing for these hybrid blends are shown in Table II and Figure 6 as a function of filler/rubber ratio. What becomes immediately obvious from this figure is that for both the 50 μm BN(5)/CTBN(5) and 50 μm BN(2.5)/CTBN(7.5) hybrid formulations the K_{IC} value falls above the ROM line indicating a positive interaction of the toughening mechanisms.

Table II. Fracture Toughness of 50 μm BN/Rubber Hybrids

Formulation	K_{IC} (MPa √m)	ΔK_{IC} (MPa √m)
Neat Epoxy	0.77±0.04	0.00
BN(5)	1.30±0.10	0.53
BN(10)	1.78±0.11	1.01
BN(5)/Rubber(5)	2.08±0.11	1.31
BN(2.5)/Rubber(7.5)	2.16±0.10	1.39
Rubber(10)	1.93±0.06	1.16

Note: Numbers in parenthesis indicate concentration of modifiers in vol. %.

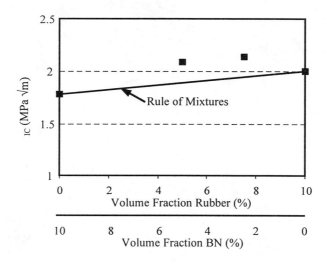

Figure 6. K_{IC} versus modifier ratio for the 50 μm BN/CTBN rubber epoxy hybrids. Note the greater than ROM toughening for all the composite ratios.

The use of transmission optical microscopy (TOM) to examine the crack tip may elucidate the synergistic toughening measured for the 50 μm BN/CTBN hybrids. TOM micrographs of the crack tip for glass sphere/rubber hybrids revealed plastic zone stretching or branching as a result of the interactions between the rigid particles and the rubber particles. For the 10 μm BN/rubber hybrids, a reduction in the plastic zone resulted in less than additive interactions of the two toughening mechanisms.

The TOM micrograph of the crack tip for the 50 μm BN(2.5)/CTBN (7.5) is shown in Figure 7. This figure shows the development of a large plastic zone in front of the crack tip. Also evident from this figure is the deflecting crack due to the BN particles interacting with the crack tip. The plastic zone tends to follow the path of the crack as it is defected. The growth of the plastic zone appears to extend out from the crack tip to the large BN particles. There does not appear to be any plastic deformation beyond these particles. From this micrograph, it is possible to see the interactions of the rubber particles and the rigid BN discs.

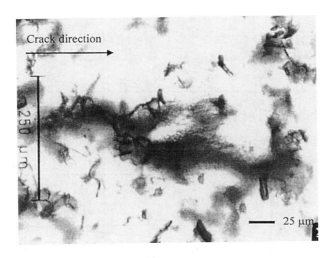

Figure 7. TOM micrograph for 50 μm BN(2.5)/CTBN(7.5) showing the development of a large plastic zone in front of the crack tip that extends out from the crack tip to the large BN particles.

The presence of the rubber particles will cause the development of a plastic zone in front of the crack tip. The plastic zone will develop outward until it encounters the rigid BN particles. The particles will shield further growth of the plastic zone by inhibiting the growth of shear bands. A similar phenomenon was observed in terms of bond line effects for rubber modified adhesive joints. Several researchers (27-30) have observed that when the bond thickness decreases below the diameter of the damage zone, the fracture resistance of the adhesive joint decreased. This was explained by the fact that the rigid adherends physically limit the plastic zone size. This suggests that when the rigid particles are closer together than the plastic zone size of the rubber modified epoxy, then the particles will physically inhibit the development of the damage zone.

To determine how the ID of the rigid particles interacts with the process zone, it will be useful to calculate the size of the rubber cavitation/damage zone. A number of authors (9,15) have shown that the fracture toughness scales with the size of this damage zone. Additionally, Irwin's equation for crack tip plasticity has been shown to provide good agreement with measured plastic zone sizes. Irwin's equation relates the plastic zone radius, r_y to the stress intensity factor, K_{IC} and the yield stress, σ_y through:

$$r_y = \frac{1}{6\pi}\left(\frac{K_{IC}^2}{\sigma_y^2}\right)$$ (3)

By using yield stress and fracture toughness data for an epoxy modified with various levels of rubber, the plastic zone radius can be determined as a function of volume fraction of rubber.

The surface-to-surface inter-particle distance (ID) for spheres can be calculated using equation 4.

$$ID = D_p\left[\left(\frac{\pi}{6V_f}\right)^{1/3} - 1\right]$$ (4)

where D_p is the particle diameter.

In order to use this equation to calculate the ID for the discs, the 10 μm and 50 μm were converted to spheres of equivalent volume. Such a calculation provides approximate particle diameters of 5 μm and 25 μm. The ID as a function of volume fraction for these BN particles along with the 50 μm spheres used by several researchers is plotted in Figure 8. This figure shows that as the volume fraction of particles decreases, the ID increases, additionally, for the smaller particle diameters i.e. 5 μm, even at the lowest volume fractions, the particles are still very close together. The condition of large number of rigid particles that are very close together may physically prevent (or shield) the growth of the shear bands.

Having established a relationship for r_y as a function of V_{f-r} and recalling the BN inter-particle distance for various filler loading levels (Figure 8), the two functions may be plotted together. Figure 9 shows a plot of ID and r_y for the 10 μm BN/CTBN and 50 μm BN/CTBN hybrid formulations (increasing V_{f-r} and decreasing V_{f-BN}). As can be seen in this figure, the ID for the 10 μm BN particles is significantly smaller than the plastic zone size for all the BN/rubber ratios studied. This results in plastic zone compression and less than additive (competitive) toughening. For the 50 μm BN particles the ID was slightly larger than the r_y enabling the process zone to develop.

Based on this analysis, the synergistic toughening observed by Azimi (9) and Smith (10) for 2.5% glass/ 7.5% rubber is due the interparticle spacing of the glass spheres being on the order of the rubber cavitation/damage zone size. Consequently, the same hybrid formulation prepared with smaller glass sphere should exhibit a less than additive increase in fracture toughness. Table III shows a comparison of the results hybrids prepared using 50 μm glass sphere and 7 μm glass spheres. As this

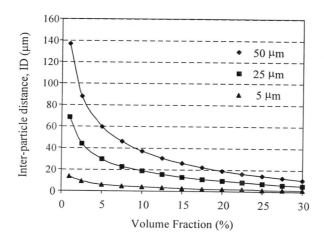

Figure 8. Plot of inter-particle distance versus volume fraction of 5, 25, and 50 μm spherical particles.

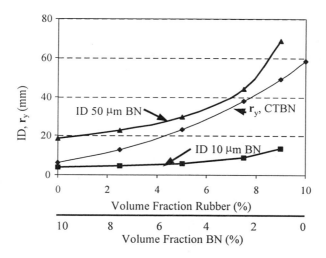

Figure 9. Plot of ID and r_y as a function of BN/rubber volume fraction ratios for the 10μm BN/CTBN and 50 μm BN/CTBN hybrid formulations.

Table III. Fracture Toughness of Glass Sphere/Rubber Hybrids

Formulation	K_{IC} (MPa √m)	ΔK_{IC} (MPa √m)
Neat Epoxy	0.90±0.02	0.00
7 μmSGS(10)	1.92±0.08	1.02
7 μmSGS(2.5)/Rubber(7.5)	2.53±0.07	1.63
50 μmSGS(10)	1.97±0.09	1.07
50 μmSGS(2.5)/Rubber(7.5)	3.54±0.09	2.64
Rubber(10)	3.20±0.04	2.30

Note: Numbers in parenthesis indicate concentration of modifiers in vol. %.

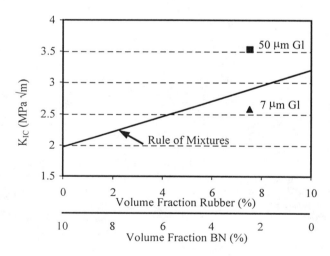

Figure 10. K_{IC} versus modifier ratio for the 7 μm and 50 μm glass sphere/CTBN rubber epoxy hybrids.

data shows, the fracture toughness for the 10% glass sphere modified epoxy is independent of particle size, although it should be noted that the toughening mechamism for the 7 μm glass spheres was found to be microcrcking. The fracture toughness for the glass sphere/rubber hybrids is dramatically higher for the large 50 μm glass spheres then the 7 μm glass sphere. These results are plotted as a function of BN-to-rubber ratio in Figure 10. This figure shows that the hybrids containing large 50 μm glass spheres exhibits synergistic toughening whereas the hybrids with the same glass/rubber ratio but with smaller 7 μm glass spheres displays less than additive toughening. This data supports the results from the BN/rubber hybrids that the interparticle spacing of the rigid particles is responsible for the presence of synergistic toughening.

Conclusions

In this study the interactions of the two toughening mechanisms associated with the rubber and BN particles were investigated. Hybrid composites were prepared by systematically varying the amount of BN and rubber particles, while maintaining a constant total volume fraction of filler particles (10%). Hybrid composites were characterized through fracture toughness as well as through microscopy. TOM microscopy was used to study the interactions of the two toughening mechanisms present in the modified epoxy systems.

The addition of small (10 μm) BN particles to a rubber modified epoxy provides an additional increase in fracture toughness over the rubber alone. The 10 μm BN/rubber hybrids exhibited fracture toughness that is less than additive for the two modifiers separately. This was shown to be due to the lack of a plastic zone in front of the crack tip for the, indicating a competition between crack path deflection associated with the BN particles and the cavitation/shear yielding from the rubber particles. The use of larger 50 μm BN particles in conjunction with the rubber particles produces hybrids that show synergistic toughening. TOM reveals a significant plastic zone resulting from rubber cavitation and epoxy shear yielding between the BN particles.

Supplemental investigations on epoxy hybrids containing glass spheres and rubber particles support the conclusion that the relation between the rubber particle induced plastic zone size and the interparticle distance between the rigid particle is a controlling factor in achieving synergistic toughening. The presence of rigid particles spaced in close proximity to the crack tip can physically inhibit or shield the development of a damage zone whereas rigid particles spaced farther apart will allow for rubber cavitation/shear yielding to occur.

References

1.   A. F. Yee and R. A. Pearson, *J. Mater. Sci.* **1986**, *21*, 2462.
2.   J. Spandoudakis and R. J. Young, *J. Mater. Sci.*, **1984**, *19*, 473.
3.   A. G. Evans, *Phil. Mag.*, **1972**, *26*, 1327.

4.   A. J. Kinloch and D. L. Hunston, *J. Mater. Sci. Lett.,* **1986**, *4*, 909.
5.   A. C. Garg and Y-W. Mai, *Composite Science and Technology*, **1988**, *31*, 179.
6.   A. J. Kinloch, D. Maxwell, and R. J. Young, *J. Mater. Sci. Lett.,* **1985**, *4*, 1276.
7.   A. J. Kinloch, D. Maxwell, and R. J. Young, *J. Mater. Sci.,* **1985**, *20*, 4169.
8.   D. Maxwell, and R. J. Young, and A. J. Kinloch, *J. Mater. Sci. Lett.,* **1984**, *3*, 9.
9.   H. R. Azimi, R. A. Pearson, and R. Hertzberg, *J. Appl. Polym. Sci.*, **1995**, *58*, 449.
10. A. K. Smith, R. A. Pearson, and A. F. Yee, *50th SPE ANTECH Meeting*, *1992*, 2631.
11. I. M Low, S. Bandyopakhyay, and Y. W. Mai, *Polym. Int.*, **1992**, *27*, 131.
12. Y. Agari, M. Tanaka, S. Nagari, and T. Uno, *J. Appl. Polym. Sci.*, **1986**, *32*, 5705.

13. A. G. Evans, Z. B. Amad, D. G. Gilbert, and P. W. R. Beamount, *Acta. Metall.*, **1986**, *34*, 79.
14. Y. Huang and A. J. Kinloch, *J. Mater. Sci,* **1992**, *27*, 2763.
15. R. A. Pearson and A. F. Yee, *J. Mater. Sci.* **1991**, *26*, 3828.
16. T. K. Chen and Y. H. Jan, *J. Mater. Sci,* **1992**, 27, 111.
17. C. K. Riew, E. H. Rowe, and A. R. Siebert, *ACS Adv. Chem. Ser.* **1976**, *154*, 326.
18. W. D. Bascom, R. Y. Ting, R. J. Moulton, C. K. Riew, and A. R. Siebert, *J. Mater. Sci,* **1981**, *16*, 2657.
19. H. Zang and L. A. Berglund, *Polym. Eng. Sci.*, **1993**, *33*, 100.
20. A. C. Rouling-Moloney, W. J. Cantwell, and H. H. Kausch, *Polym. Composites*, **1987**, *8*, 314.
21. G. Vekinis, P. W. R. Beaumont, G. Pritchard, and R. Wainwright, *J. Mater. Sci.,* **1991**, *26*, 4527.
22. C. I. Vallo, Lisiang Hu, P. M. Frontini, and R. J. J. Williams, *J. Mater. Sci.,* **1994**, *29*, 2481.
23. K. T. Faber and A.G. Evans, *Acta. Metall.*, **1983**, *31*, 565.
24. M. F. DiBerardino and R. A. Pearson, *55th SPE ANTECH Meeting*, **1997**.
25. M. F. DiBerardino and R. A. Pearson, *Mat. Res. Symp. Proc.*, **1998**, *515*, 239.
26. *Annual Book of ASTM Standards*; Standard D5045; American Society for Testing and Materials: Philadelphia, PA 1991.
27. W. D. Bascom, R. L. Cottington, R. L. Jones, and P. Peyer, *J. Appl Polym. Sci.*, **1975**, *19*, 2545.
28. W. D. Bascom and R. L. Cottington, *J. Adhesion*, **1976**, *7*, 333.
29. A. J. Kinloch and S. Shaw, *J. Adhesion*, **1981**, *12*, 59.
30. S. S. Wang, J. F. Mandell, and F. J. McGarry, *Int. J. Fract.*, **1978**, *14*, 36.

INDEXES

Author Index

Subject Index